My Dear Husband

COMPILED AND EDITED BY
GENEVIEVE M. DARDEN

A Project of
The Descendants of Whaling Masters, Inc.

LCCCNo. 80-50605
ISBN 0-88492-036-4

William S. Sullwold Publishing, Inc.
Taunton, Massachusetts

FOREWORD

"A major experiment in the illusions of fiction was the use of letters, a method more effective and brilliant than today's writers quite realize. The difficulties of it are formidable, the demands on our belief are great, the dramatic moment is awkward, but the effect of immediate vitality can be superb .... A letter is an immediate event; it is written in the very midst of experience."

Gerald Warner Brace 1901-1978
The Stuff of Fiction
W.W. Norton and Co., Inc., 1969

This thought from a grand old New England novelist, who knew and loved the sea as thoroughly as he understood and cherished his profession, has been close to my mind as I've put together this collection of hitherto-unpublished letters of the whaling era. While there is neither illusion nor fiction in what follows, because I have taken only such liberties as would eliminate repetition and place each family in some kind of perspective for the reader, I have become more aware of the contents' "vitality" with each re-reading. Letters are indeed immediate events, "written in the very midst of experience." Therefore each series, as each contributing member of the Descendants of Whaling Masters has realized, gives us a bit more insight to the human side of a bold, colorful --- but lonely and exacting --- time in New England's history.

Genevieve M. Darden

PREFACE

In the pages that follow, the human qualities that emerge most frequently are loneliness, isolation and the patient endurance of long absence. Sometimes disguised by enforced cheerfulness, sometimes displaced by tales of faraway adventure, the emotions aroused by separation can be discerned in almost every line of every letter. Indeed, we should hardly expect otherwise, for the lives of those who served in the whaling trade were marked by recurring departures, lengthy voyages and only brief periods at home.

The central predicament of the whaleman and his wife once attracted the attention of the most profound of all whaling writers. The episode is pertinent and worth recalling. In 1853, Herman Melville passed through New Bedford on his way to Nantucket. A prominent local lawyer, John H. Clifford, joined him for the journey and told him of a case he had been involved in years before. A Cape Cod woman, Agatha Robertson, had been deserted by her sailor husband after two years of marriage. Suddenly, the husband returned and offered to resume the relationship. Although he omitted to tell her that he had married again and had a wife in another port, Agatha accepted. Later, however, she declined to accede to his wish that they move to Missouri. He was adamant, so she offered neither objection nor reproach to his proposal that he marry someone else, in order to have a companion in the West.

To Melville, it was a "story of remarkable interest." Unsure of his ability to do it justice, he passed it on to Nathaniel Hawthorne: it lies, he said, "very much in a vein, with which you are peculiarly familiar." In the end, neither Melville nor Hawthorne transformed the melancholy tale into a work of literature, although both were quick to see the possibilities of a story based on "the peculiarly latitudinarian notions which most sailors have of all tender obligations" and the "resignedness of the women in submitting so uncomplainingly to the long, long absences of their sailor husbands."

Most of the letters in this book reflect stronger ties of devotion rather than the Robertsons' latitudinarian notions. What we have here, in fact, is the other side of the coin: the written efforts, however strained, of whalemen and wives to bridge the separation of time and distance by the only means available to them. Their letters stand as testimonials to the endurance of the human heart, and are as much worth reading on that account as for the information they contain on the actual business of whaling.

Richard C. Kugler
Director, New Bedford Whaling Museum
DWM Honorary Member

Captain and Mrs. George Lyman Howland are shown with their sons Chester Scott, left, and James Lyman, right. Chester was to become a prolific writer of whaling tales, and James the father of Ellen Howland, first president of the Descendants of Whaling Masters. Captain Howland was master for many voyages of the Canton.

ACKNOWLEDGMENTS

Letters, journal excerpts, and related materials which appear in this volume have been contributed by the following members and friends of the Descendants of Whaling Masters:

Captain James Ryder Allen	Gladys Hersom and the late Allen H. Hersom, South Dartmouth, Massachusetts
First Mate William A. Ashley	Sophronia Howard, Acushnet, Massachusetts
Captain Seth Mendell Blackmer	Winifred B. Reynolds, Wilbraham, Massachusetts
Captain Paul Cook Burgess	Genevieve M. Darden, Fairhaven, Massachusetts
Seaman Benjamin Franklin Gifford	Grace M. and Morgan Levine, Palm Beach Gardens, Florida
Captain Charles Pierce	Helen H. Radcliffe (letters) and Kenneth S. Peirce (illustration) Fairhaven, Massachusetts
Captain Francis Post	Ruth P. St. Lawrence, Tiverton, Rhode Island, and Irving R. Post, (portrait), Sharon, Massachusetts
Captain Sheffel Read	The late Clara B. Marston, Fairhaven, Massachusetts
Captain James Henry Sherman	Ruby and Carolyn Sherman (letters) and Ellen Howland (illustrations) New Bedford, Massachusetts

Captain Albert A. Thomas	Sylvia Thomas, Flushing, New York
Captain Rodolphus Delano Wicks	Virginia A. Taylor, New Bedford, Massachusetts, and Helen Bean, Concord, New Hampshire

Thanks in abundance are also expressed to Ellen Howland, president of DWM, and Dorothy Howland, an honorary member, for consistent and constructive help during the two years required to assemble the letters which follow; to Richard C. Kugler, director of the Whaling Museum of the Old Dartmouth Historical Society; Philip F. Purrington, senior curator of the museum, and Nicholas Whitman, museum photographer; for the resources of the Kendall Whaling Museum, Sharon, Massachusetts, and those of the Nantucket Whaling Museum, Nantucket Historical Association, Nantucket, Massachusetts; to the Rev. Robert Hardy of Acushnet, Massachusetts; to patient staff members of town and city clerks' offices in Acushnet, Fairhaven, and New Bedford, Massachusetts, and to equally helpful personnel of the Millicent Library, Fairhaven, and the Mattapoisett Public Library, Mattapoisett, Massachusetts. We are also grateful to the many persons who undertook the time-consuming, meticulous task of transcribing 19th Century handwriting.

This Albert Cook Church photo of the Canton, outward bound, is owned by the New Bedford Whaling Museum.

CONTENTS

The Eagle may have looked as triumphant as this when she escaped, under command of Captain James Ryder Allen, the Confederate marauder Alabama in 1862. Photo courtesy of the New Bedford Whaling Museum.

JAMES RYDER ALLEN

"As for the Pirates I would not mind any thing about them in your place... I should cruise and do just the same as though I never heard of one and should I meet one if I could not run away I'd run him down if I could."

"You must have had a pretty hard time, with bad weather, parting fluke chains, and being chased by the pirate Alabama... You were indeed lucky to escape him, but we think he cheated us out of considerable oil as whales were quite plenty late in the season."

So wrote Swift and Perry of New Bedford, a pair of down-to-earth managing agents, to Captain James Ryder Allen, who had been harried and "fired upon three times" by the Confederate vessel Alabama, under her ferocious Admiral Raphael Semmes, several months before.

Allen saved the bark Eagle from Semmes, but the southerners drove him in to Flores, in the Azores, where in July of 1862 he found ninety-three New England whalemen who had not been so lucky. Their vessels reduced to cinders by the Alabama, they had despaired of ever getting home again until Captain Allen, "under cover of darkness" and some eight days after his arrival, smuggled them out to the larger Azores port of Fayal, where they would be able to find passage back to the United States.

"Yours of no date from Fayal came safe to hand," commented Swift and Perry from New Bedford. "Be assured we was all well pleased with your good success but we all feel that that pirate who was about the isles done you out of as much more... Oil sold yesterday for 1.75 per gal and no doubt will go higher.

"Your oil came home in fine order and was splendid in coulour but weighed somewhat heavy which was verry well for purchasers."

So much for the travail which Allen, and dozens of other New England masters of commercial ships, survived or succumbed to during the turbulence of the Civil War.

The bark Eagle was one of six whaling vessels which Allen, who had gone to sea when he was barely thirteen, commanded between 1848 and 1865. Probably his most famous charge was the Josephine---"the prettiest whaling ship ever to sail out of New Bedford"---built in 1856 in Fairhaven, expressly for him, by Swift and Perry. He took her 446 tons to the Pacific on her maiden voyage in July of 1856, to return thirty-three months later with a satisfying

"157 barrels sperm oil, 3,371 barrels whale oil, 37,000 pounds whale bone" aboard.

He also had other pleasing things on his mind. Toward the end of that October, 1859, he took as his second wife Pamelia Cowen Douglas of Rochester. (Abigail Bennett Allen had died eight years before, at the age of twenty-six.) The captain was then forty-three; the arrival of a daughter a year later delighted him, so much so that he reversed the usual order of nomenclature by naming her for his ship --- she became Josephine Chapman Allen. (The middle name honored his "esteemed friend and old shipmate" James L. Chapman---master of the Josephine for two voyages after Allen--- who recriprocated in 1872 by calling his newborn son Leslie Allen Chapman.)

"I must write a few lines to congratulate you on the acquisition of an heir, but I expect it would have been more pleasing had it been a son but the probability is it will be all right yet," wrote George M. Stetson of Camden, Delaware, to Allen in October of 1860.

"Of course this will settle the question of your going to sea again as you would not think for a moment of leaveing so much comfort and enjoyment as you are at present possessed of and you would not think of taking your wife and child (how queer that sounds, dont it) to buffet the storms of Old Ocean to no purpose whatever only for persons to say after you are laid under the sod that he left a few thousands. . . ."

Stetson was right about his old colleague staying in from the sea, but only for two years. On June 17, 1862, Captain Allen once again left New Bedford, on the bark Eagle, to begin the voyage which led him into the path of Semmes and the Alabama.

"You will no doubt want to hear something about the war," reported O.N. Swift, a retired and ailing New Bedford master, in an unconsciously-innocent letter which he sent out to Allen in Fayal the following February. "Well my dear fellow it drags along and we do not get along well at all and all hope something may be done.

"As for the Pirates I would not mind any thing about them in your place as they will hardly come out into the Rugged country to cruise, if by any chance you should see one he would not burn your ketchings up. I do not know what Messrs Swift and Perry think of it they no doubt will write you their views on the subject. I should cruise and do just the same as though I never heard of one and should I meet one if I could not run away I'd run him down if I could, so I hope you will dismiss from your mind all fears on their account.

"My wife has left for your house to see Mrs. Allen and the little one so that I can tell you how they are also to tell her the Hortense is in Boston and will sail for the isles soon and she had best write to you.

"Frank Nye left as a Soldier last fall is now in North Carolina, a number of our young men went last fall but they do not fite much yet, it takes them all the time to get ready.

The port of Fayal, Azores, in the 1860's, as depicted by a section of the Purrington and Russell Panorama photographed and lent by the New Bedford Whaling Museum.

"the doct has just called and says I must not write to much as it is to much for me as you will see by my writing that I am quite nervous and my hand quite unsteady but so that you can read it I do not care."

The news of the Alabama's attack on the Eagle seven months before had obviously not reached old Captain Swift; the canny agents Swift and Perry had known of Allen's close call for several weeks, however, for in late February of 1863 they wrote:

"Our Government has taken vigorous measures to catch the Alabama and other southern privateers and we hope that the Atlantic Ocean will be rid of them before you come North this summer. We should hardly know where to advise you to cruise either as regards safety from privateers or finding whales." And then, ever practical, and having noted in an earlier letter the burning of the whaler Levi Starbuck, "We think it would be best to ship what oil you have taken this Winter as you can probably do so from Fayal in foreign bottoms, or leave it there, to be shipped in safety.

"In our last letter we sent you an extract from Capt. Dexter's letter, but as you may possibly not get it will repeat it in this:

"... on the last day of September, Lat. 40.42 Long 39.30 he saw 6 large whales going NNW and kept them in sight 5 days. He says there is a small spot of Green Water in Lat. 42.20 Long 41.18 where all these whales stop...

"We do not think it would be of much use to try to keep out of the tracks of any privateers that might be cruising in the Atlantic. We are in hopes that our Government may succeed in catching any that may be out but it is possible they may not. Wishing you health and a prosperous voyage ...."

It's hard to avoid resurrecting the French axiom that "the more things change, the more they remain the same." As it is now, so it was then. Good luck outwitting Semmes, Allen, and save your own skin if you can, but deliver the oil.

No letters indicate how the skipper of the Eagle took his agents' advice; probably he was as matter-of-fact about his duty as they were. In any event, pirates or no pirates, and as if nothing untoward had happened, he went on cruising. In late December he lost second mate Silas B. Plato and an entire boat crew while they were "fast to a whale" somewhere in the Atlantic but such tragedies, as whalemen knew, were all part of the business.

Captain Allen returned to New Bedford, his ship reasonably full, on October 7, 1864, and shortly thereafter Swift and Perry recorded complacently, also part of the business, that they got "$1.70 to $1.75 per Gallon for the Oil and $1.30 to $1.50 per Pound for the Bone."

On the subject of figures it's perhaps of topical interest to note that initial provisions for the Eagle's trip included "8½ lbs coffee, $1.53; 1½ lbs pepper, .25; 1½ lbs. starch, .18; 2 gals Molasses, .96; 7¾ lbs lard, $1.00; 2 bags Salt, .16; 1 lb Ginger, .12, and 6 lbs. Saleratus, .51."

And on the subject of human beings in a particular time in history, which is what this book hopes to be about: It is no secret that whaling captains were often as tough as the hides of the whales they pursued, and frequently---in the light of much we now know, or at least suspect, about the giant mammals' intelligence and dispositions---a dozen times more brutal. Allen seems to have been of a different stripe. As early as 1853 Swift and Perry had chided him, in a letter addressed to the whaler Pacific, for allowing "your ship's company to draw an exceedingly large percentage of the profits.

"We have full confidence that you will take all the pains in your power to do the thing that is right by us and make all the savings for the ship you can consistently make. Of course you will ship your bone. Sell oil rather than draw."

How much attention Allen paid to this admonition is not recorded, but the suggestion that he went on in his own relatively compassionate fashion emerges from some otherwise-enigmatic verses written in the same year, 1853, and addressed to him from "Santa cruise":

In bondage thou didst find me
The iron bands did see
And when the truth was told thee said
I'll set my brother free,

It was said and done thy heart was true
And nobly didst thou brave
Soon from the land of tirany
Thou didst the victim save.

One stanza precedes the above, and six more follow. While it is true that whalemen indulged fairly often in lengthy iambics, many of which were dreadful (it was after all the age of Victoria, who managed to make her influence felt in most of the world's pockets) the above writer at least had something substantial to say. No one seems to know the genesis of the incident, the reminder of which the captain abviously treasured enough to save, but the message ends:

Accept these lines, my loving friend
May they never be erased
They come from one who holds thee dear
Thy Brother George H. Chase.

Allen's last voyage was as master of the Pacific, out of New Bedford, in 1865. He had been captain of the Condor, Canton, Pacific, Josephine, Eagle, and Pacific again, in that order, had sailed on the first American whaling ship to visit China, and died on December 6, 1904.

"Captain James R. Allen, one of the crack whalemen of his day, died today at his home in Acushnet, aged 88 years, 5 months," The Evening Standard of New Bedford reported.

"He was born in South Dartmouth July 30, 1816, and when but 13 years of age shipped for his first voyage, sailing in the old ship Columbus of Fairhaven. His second venture was in the Marcia of Fairhaven, Captain Peter Butler Jr. . . . Captain Allen advanced by degrees, sailing in the America, Adeline and Uncas until July 7, 1846, when he went out as mate on the Condor with Captain Jacob Taber, and then on June 23, 1848, three months home, he was given his first command, which was the Condor. This was his eighth voyage whaling and he was then but 30 years of age . . ."

Even a cursory and somewhat patchy account of James Ryder Allen's career---the lack of letters written by the captain himself is regrettable---requires a concluding note about the Josephine. In the 1860's and 1870's she left New Bedford, for about four years each voyage, under Captains Chapman, Bernard Cogan, and George F. Long. Then in the 1870's, commanded by Charles F. Hamill, she was caught in that now-famous North Pacific ice, "in common with a large group of other whalers" and abandoned. She survived, however, perhaps because of her stout heart and build, eventually made it back to New Bedford, was re-rigged as a bark, and in September of 1909, five years after the death of her first master, was sold by the J. and W.R. Wing Company to Compania Cheneira de Balleneros of Valparaiso, Chile, which planned to sail her out of that port after sperm and right whales.

The bark Josephine at a New Bedford wharf in 1905. Courtesy of the New Bedford Whaling Museum.

WILLIAM A. TO HANNAH H. ASHLEY

". . . the ship was in ice sose we could not handle her and going around the north pole like a streak."

"I hear by some of the folks that is right out from home that the Ladies have got to wareing tailes. If you ware one look out there that you do not get trod upon. I should like first rate to be at home and see the Ladies tailes hanging out behind. you must make allowance for Sailors, that is so."

So wrote William A. Ashley, first mate on the whaler Governor Troup, to his wife Hannah in 1864. The Troup was "about as far north as a ship can get," and within two weeks of being jammed in ice when young Ashley, who managed to keep up a "faint lively hart" in plights as dismal as most whalemen would care to imagine, began on June 15:

"In the ice blowing like fun and no whales. My dear wife Hannah A. Ashley we are all well at present and hope and pray that you are so to . . . I think sometimes if I ever get home again alive I will never leave you again.

"Oh what is the use to talk all this. I want to see you awfully and I cant help it. We have had rather hard luck so far, havent seen a whale since we left the islands but never mind. I do not fret a mite about it. I only say let her rip. I think more about you than anything else how you are getting along out there to Long Plain with a Baby and all those scalawag things to look out for."

Hannah Crapo had been married to William Ashley of Acushnet, near the Rochester line---where some of their descendants still live---on July 4, 1860. On a cold morning in early December, 1862, William shipped out of New Bedford on the Troup as next in command to his brother, Captain Edward R. Ashley, "the old man" of his letters.

"We have had very bad thick weather here for the week past," he wrote Hannah on June 17, 1864. "We have not seen any whales yet but we have got up whare they ort to be for it is cold enough and ice around enough . . . Oh Han I am getting real homesick and am in hopes that we shall get enough oil this season to come home if we do not I do not no but I shall run away or do something despret. Kiss Willie for me."

Willie, the Ashleys' only child, was then about fourteen months old. In a diary kept during the year of his birth, 1863, his mother reveals the flintiness for which she was to become regionally famous when she tells, on March 7, of "whitewashing the kitchen," and then writes, on March 8:

"Today felt some pain. Our son William was born this evening."

Willie's father's reaction to this event is missing, but we do know that he spent his next wedding anniversary, the fourth, somewhere in the Arctic Ocean, where he wrote his wife on July 5:

"here we are tied up to the ice and have been for the last week. I have been thinking of you a great deal lately as it was about the forth of July in the larst letter you wrote to me. You asked me if I remembered three years ago then and now I ask you if you remember 4 years ago last night.

"last night to me was a night never to be forgotten. the ship was in ice sose we could not handle her and going around the north pole like a streak, we was pretty well north when we got jamed in it but it is brighter now for getting out but I do not no as we might as well be around the north pole as any whare else if we do not get any oil we have not got any yet and do not no when we shall but we are all tuff and harty and that is the most I think of . . . There is a great many that have been taken away by death this season.

"we have seen some whales since we have been in the ice but could not catch them the ice is so thick but we are in hopes to find it thinner by and by and the whales thicker."

The frustration of being "in the ice" is reflected ten days later in "all the Ships up here has got a good cut while we were jamed in the Ice and the current taking us around the North Pole but thank fortune we have got out of that scrape in good shape only we left some of our copper in those reagons.

"you need not make much talk about our being jamed in ice for it might be the cause of considerable talk. I am bound to keep up a faint lively hart and you must try to do the same oil or no oil. give my love to all and tell them I am as smart as a deer and can jump higher. Goodbye dear Wife for I am always thinking of you take care of that billy boy."

Toward the end of July, still "nary oil yet," he sends the even more somber report that "the small pox has got in the fleet and is raging pretty bad on board some of the Ships, there has been quite a number died of it," but by September 2, back in his old form:

"All well and kicking. there is now a Ship coming down that we think is bound to the Islands and thinking there will be a chance to send a letter I will scratch a few lines to send with that other I have ready. We have got six whales that made us about 500 bls but I think there is a chance for more if we stay long enough, but it is a very hard season and I believe that hard luck follows us wherever we go but never mind I am tuff and harty.

"Dear you must keep up good spirits and not get down harted for we shall come out all right in the end, you must not never build your hopes to high on our seasons work and then you will not get disapointed. . . Han take good care of the baby boy and be a good girl and do not get into any mischief. Tell Father and Mother that I am as tuff as a shark and if I was at home I would make that old salt meadow ring."

From the Sea of Okhotsk, which balloons out of the North Pacific, he added on September 17:

"the dogs are going with a clean rush today and I thort I would rush a few lines to you. . . this has been the hardest season ever known at sea, smock and fog all the time and very few whales. the old mans had very sore eyes but got over it now, he sends his love to you and says you must keep your nose clean.

"we shall probably be in Hilo (Hawaii) in two months and I shall expect a heap of letters from you. han you must be a good gall and take good care of yourself for I am out there to help you. (Not only did Hannah, who lived to be a "peppery old lady of almost a hundred,"[1] take good care of herself, but she apparently did as much work on the Ashley farm during William's absence as any two men. She and her horse Thad, whom she rode regularly to Lunds Corner, New Bedford — taking care to change to clean gloves before she began her round of farm and domestic errands — became as routine and respected a sight over the years as the old church steeple.) All the Ships have done very poorly this season. . . Capt. Willice is clean and had a boats crew run away and caried three bomb guns dont think he will catch them.

"I shall expect your likeness when we get to the Islands.

"We came in here for the purpose to bury a man, a Mr. Cranston, he was the old mans mate in the William Wert, he shipped with us for 5 mate but was sick when we left the Islands he leaves a wife at home but no little ones thank fortune.

"Keep up good courage dear Han for one third of the voige has gone allready if you want anything dont you fail to rite me and you shall have it if in the bounds of reason. I am always thinking of you."

On September 22 "safe and sound through the Straits on the old Right Whale ground" [the Kodiak area, in the Gulf of Alaska] he gives the advice about what to do should she wear one of the new "Ladies tailes" and from here on, with the exception of one 20th Century observation, it seems appropriate to let a master stylist speak for himself:

Lat 51 Lon 170
Old Right Whale Ground
September 23

Much Beloved Wife as I have a few spare moments I will imploy them writing to you. Han I am getting real lonesome lately and I begin to feel as though it would help me if I could see you for a few minutes. what do you think. now dont you think it would. now what am I to do. I do grin and bare it. well that is hard but fair I spose, well enough of this gass.

---

[1]The Rev. Robert Hardy, husband of an Ashley descendant, Acushnet, Massachusetts.

This Benjamin Russell painting of the whaler Governor Troup is part of the Hart Nautical Museum's Forbes Collection, Massachusetts Institute of Technology, which has given permission for its reproduction here.

I expect you are disapointed when you come to hear that I am going another season, but if you are not I am very much so for I was in hopes that we should get home off of two seasons, it does very well for a single man to be gone a life time but I do not think it is healthy for the married ones. . . Tell Father and Mother to keep up good courage and not to give up the Ship for I shall come on board by and by and then we will make them stand around there.

I believe I spoke about the Small Pox being in the Whaling fleet. I have not heard of its being in any more than three ships — the Barque Abigail, the Jiry Swift, and Camilla. they lost 5 or 6 men apiece. you get Willie vacinated the first opportunity.

(Dr. Edward Jenner's pioneer inoculation of a British farm boy against smallpox, using infectious materials from lesions of cowpox, had been a success in 1798, and in 1864 was still the only recognized form of immunization in Europe and the United States. Why seamen were not routinely vaccinated, when they were so often victims of the disease, is matter for speculation.)

September 25

Good evening all. It is about milkin time here but bed time there and I suppose you are all snoring away like fun. I can see you and the baby laying off and taking it easy, I hope so and I wish I was there with you but time is wareing away and I shall be there before you think of it. ding dong goes the bell that calls Mr. Ashley for four hours so good night dear Hannah.

October the 16

We have got down in to warm weather Lat 23 Lon 152 cloast to the Sandwich Is. I did not have a chance to send this before but I will send it now in the Brig Rozyleen she is as pretty as her name.

It is a dead calm here and has been for several days. I never saw such weather in my life in these latitudes no wind nor never expect to get any more, if we do we shall give the old Gov. fits to get in and get the nuse and something fresh to eat.

Han I have been thinking it is too bad to keep you there to Fathers working like a squaw so I want you to hire a good girl to help you and then if you can not get along move to New Bedford and write me and I will send you a way to live if you have not got enough. I shall go one more Season in the Ship and then home oil or no oil. I do not know but the Ship will go 4 seasons but let her go I am not.

If those coalts get unruly sell them or if you think they are unprofitable. take good care of the boy. Tell Father and Mother that I shall be there in 18 months and think about it yourself a little. goodby dear Wife.

The Governor Troup did not return to New Bedford until June, 1867, four and a half years after she had cleared that port in December of 1862. Perhaps it's ironic that William Ashley, then twenty-nine, had shipped home from Honolulu in 1865—as he had sworn to Hannah he would, "oil or no oil" —for the Troup was full to the brim and running over when she finally sailed up the harbor. And then again he may well have felt that "it all comes out right in the end."

In any event, there is no record of his going whaling out of New Bedford, or any other port, again. As subsequent Acushnet directories list a "William A. Ashley, farmer, Long Plain," it seems probable that he stayed close to Hannah, Willie, and the old salt meadow, and that the voyage aboard the Troup was his last.

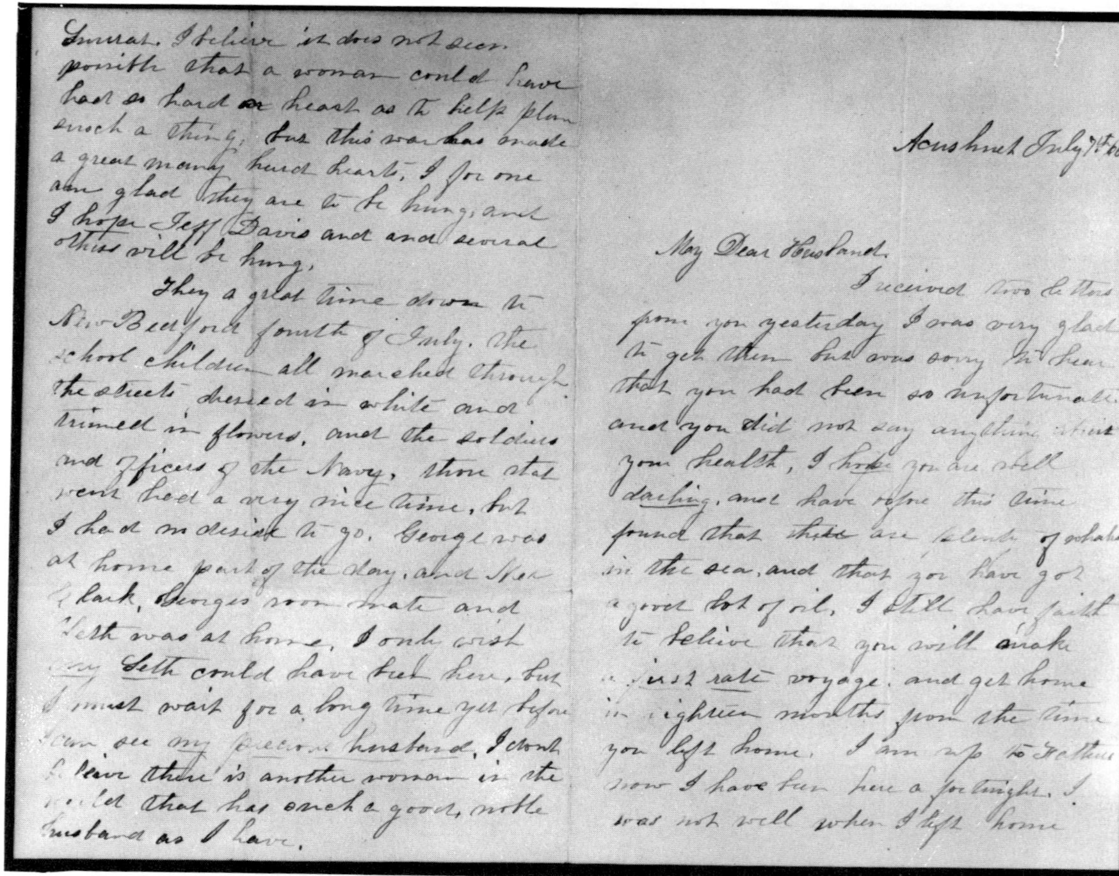
Acushnet July 7th [65]

My Dear Husband,

I received two letters from you yesterday I was very glad to get them but was sorry to hear that you had been so unfortunate and you did not say anything about your health, I hope you are still darling, and have before this time found that there are plenty of whales in the sea, and that you have got a good lot of oil, I still have faith to believe that you will make a first rate voyage, and get home in eighteen months from the time you left home. I am up to Fathers now I have been here a fortnight. I was not well when I left home

Surrat. I believe it does not seem possible that a woman could have had so hard a heart as to help plan such a thing, but this war has made a great many hard hearts. I for one am glad they are to be hung, and I hope Jeff Davis and and several others will be hung.

They a great time down to New Bedford fourth of July. the school children all marched through the streets dressed in white and trimmed in flowers, and the soldiers and officers of the Navy, those that went had a very nice time, but I had no desire to go. George was at home part of the day, and Stev Clark, Georges room mate and Seth was at home. I only wish my Seth could have been here, but I must wait for a long time yet before I can see my precious husband. I don't believe there is another woman in the world that has such a good, noble husband as I have.

"I for one am glad they are to be hung, and I hope Jeff Davis and several others will be hung." So wrote Hannah Mendell Blackmer of Acushnet to her husband, Captain Seth M. Blackmer, on the historic date of July 7, 1865.

HANNAH AND SETH MENDELL BLACKMER

"Father says tell Seth that four or five of the assassins are to be hung today, and one of them is a woman, Mrs. Surrat. . . I for one am glad they are to be hung, and I hope Jeff Davis and several others will be hung."

"We have conquered the South, the Confederate armies have surrendered to Generals Grant and Sherman. Jeff Davis and staff have been beaten. They are now being carried to Washington. Davis will be hung. President Lincoln was shot and killed while attending the theater by an assassin named Booth. Secretary Seward's life was attempted at the same time at his house. He was twice stabbed but is fast recovering. We have most of the assassin's accomplices who are now being tried and will be hung."

The letter above is dated May 20, 1865, and addressed to Captain Seth Mendell Blackmer, Bark Stella, St. Helena---the mountainous island in the South Atlantic to which Napoleon had been exiled fifty years before. The writer was Blackmer's New Bedford agent Loum Snow.

"The death of President Lincoln has not stopped for a moment any department in our government," Snow adds. "This has shown our Republic stronger and firmer than all other forms of government and we stand before the world stronger and more respected than ever before."

Then apparently fearing that the message would not reach its destination, as happened so frequently in an era in which conclusiveness of any kind often depended upon chance, Snow and Son wrote again on May 24:

"The Rebel armies have surrendered. Jeff Davis and staff are caught and will be tried for treason and accomplice of the assassins and hung. Abraham Lincoln was shot by 'Booth' an assassin while attending a play at the theater. Sec. Seward's life was attempted the same night when sick at his house he was twice stabbed but is fast recovering. Booth was shot in being captured. . . Andrew Johnson vice President took the Oath as President and all other departments went on without the least interruption. We all feel proud and rejoice at the glorious triumph of our arms and flag.

"Gold has declined rapidly since we have put down the rebellion. Now sells 1.30. With the decline in gold all descriptions of merchandise has fallen...

Oils have fallen less than most other articles, and we think will continue to sell at about present value which is sperm 2.00, whale 1.20, whalebone 1.20 to 1.42. Bone will sell at gold rates say at about 95¢ lb. gold as nearly all the bone goes out of the country. Sperm oil will, we think, continue to sell at present price, as very few ships are continuing sperm whaling.

"The whaling fleet continues to decrease, now in port for sale Ships St. George, Desdemona, Swallow, Trident, Atlanta, Courser, Pioneer, Stafford, Mermaid, John Wells, and Norman. . . ."

War and related news was as slow to reach voyaging whalemen as were letters; newspapers were often four or five months old when they caught up with vessels, and scantily informed at that. In the autumn of 1865, therefore, Captain Blackmer--whose chief contact with the conflict had been several "lucky escapes from rebel cruisers" which stalked Northern ships — no doubt welcomed even more than usual a letter from his wife Hannah. It was written from Acushnet, Massachusetts, on the historic date of July 7, 1865:

"Father says tell Seth that four or five of the assassins are to be hung to-day, and one of them is a woman, Mrs. Surrat. I believe it does not seem possible that a woman could have been so hard of heart as to help plan such a thing, but this war has made a great many hard hearts. I for one am glad they are to be hung, and I hope Jeff Davis and several others will be hung."

(Mary Surratt, who probably had little or nothing to do with the assassination plot and whose execution is now regarded as a "gross miscarriage of justice," was indeed hanged, with three others, on July 7, 1865. Jefferson Davis, though indicted in 1866, never came to trial.)

Mrs. Blackmer's topical commentary is followed by a host of domestic details, including the complaint that "It is almost impossible to get anyone to *help* in the kitchen now. Those you can get want to eat up and play lady, and let the mistress do the work."

And then on a more positive note,

"They had a great time down to New Bedford fourth of July. The school children all marched through the streets dressed in white and trimmed in flowers, and the soldiers and officers of the Navy were there. Those that went had a very nice time, but I had no desire to go. George was at home part of the day, and Seth. I only wish *my Seth* could have been here, but I must wait for a long time yet before I see my precious husband. I don't think there is another woman in the world that has such a good, noble husband as I have."

Hannah Blackmer was the captain's second wife, and she bore him three sons and four daughters. The large family was a joy to him, for his own boyhood had been lonely. Born in 1819 in what was then Acushnet, but is now part of the neighboring town of Fairhaven, Massachusetts, he was an orphan by the time he was six. His mother, Lucy Ann Mendell, died at twenty-six and his father, Salisbury, of Rochester, who ran a tavern in his home on the road to Long Plain and sired his three children there, outlived his wife by only four years.

Seth was taken in for awhile by his grandfather, Ellis Mendell, and later by the Uncle Seth Mendell of Mattapoisett for whom he was named. With this uncle he worked the family farm during the summers and went to school winters until he was fifteen—that magical age at which a yen for the sea so often manifested itself in boys along New England's coast.

Perhaps because Seth Mendell was in poor health, and wanted to be sure that his nephew could make a living, he let the young Seth Blackmer sign on as cabin boy aboard the whaler Pacific in 1834, under Captain David Collins. The uncle's premonition about himself was accurate; he died before the Pacific returned to New Bedford thirty months later, and Seth Blackmer was once more alone.

The captain did not mention this loss in an informal story of his life which he gave to New Bedford reporter Frederick A. Bradford in 1890, however; he was no whiner, and Bradford's quotes, accompanied by a picture of his subject, indicate that the whaleman -- who made ten voyages before he retired from the sea in 1871—had a spirit both resilient and tough.

His chief memory of that first early voyage, as a matter of fact, aside from witnessing the fringe of a rebellion in Chile and seeing Captain Collins give reluctant asylum to three of the insurgents, seems to have been that "the try pots got overheated while we were trying out one day and came near setting the vessel on fire. As it was it damaged the rigging somewhat and gave us a pretty good fright.

"We got about 2600 sperm and whale oil together. I got the 175th lay" (or share of profits in lieu of wages).

Seth liked the vessel Pacific well enough to make a second cruise, which lasted about two years, in her, and by that time he must have decided that he was a whaleman for life, for he was barely back in New Bedford when he left again, in 1840, aboard the Montpelier.

His first trip as master came in 1845, when he was twenty-six; the ship was the Julian, followed by the Roman, which he skippered to the coast of Siberia and the Japanese Sea after bowheads and, unwittingly, his first typhoon.

"It was in the winter of 1848 in the West Fiji Islands," he told Bradford. "A book by Kreuzenstein that told of the premonitions of a typhoon said that the first day would be very calm, the second day would assume a leaden hue, the third day the sun would become totally obscured, and then it was time to look out.

"The symptoms about us corresponded with the book exactly, so I told the mate to secure everything. . . about 7 o'clock that night the storm struck us. I was sailing on the port tack, and the boats on the cranes to windward were struck with such force that they were stove. They were not struck by the water at all; simply the wind.

"The gale blew for six hours. It sounded like the booming of a cannon, a constant roar. A Russia canvas mainsail was blown from the gaskets and torn to shreds. It seemed to me that daylight would never come. Everything was swept from the deck that was not lashed down and the men were fast to the life lines to prevent their going overboard.

"I had thought the gale could blow no harder, but later the wind changed and blew still harder for six hours more. Then we shipped a big sea that washed everything off the deck that the wind had left."

The Roman survived although, in the understatement more or less typical of a whaling master (except when he was homesick) Blackmer commented only that she was "rather the worse for wear."

In June of 1851 "having lived alone long enough," Seth Blackmer married his eighteen-year-old cousin Catherine Mendell of Acushnet, and stayed ashore until October. In that month he sailed from New Bedford for the seventh time, bound for the Arctic aboard the whaler Erie.

Catherine was probably ill when he left, though there are no letters to confirm this assumption. When he returned in the spring of 1854, however, she had been dead for almost a year. Town records list "consumption," the 19th Century's dreaded killer of the young--along with meningitis, which was generally called some kind of "fever" --as the cause.

Three years passed before Seth, on June 27, 1857, married Catherine's sister Hannah Hammond Mendell, who was seventeen. Hannah was to live until 1913, ten years longer than Seth, and she was apparently an ideal wife.

"I am so happy tonight, dear Seth, for I have had three letters from you this week," she wrote on November 17, 1864. "It seems so nice to have you tell me what to do, for I don't think I understand much about outdoor affairs. It is almost impossible to get anyone to do a thing. Rodolphus will not come without 'Timely warning' he says, and that is three or four weeks before I want him. I don't think I should have got my corn husked this winter if I had not had that husking. . . The turnips are very nice and very large. I have sold two bushel to Jacob for sixty cents a bushel.

"I have a very good girl indeed, but she wants every Sunday and one day in the week. She goes home every Thursday and does not get back until nine

Captain Seth Mendell Blackmer made ten whaling voyages out of New Bedford before he retired from the sea in 1871.

o'clock in the evening, but then I have a nice time to write to you, for I cannot have the children round when I write. Freddie wants very much that I should go on the water to get Papa. The dear little fellow does not realize how long I should have to be gone before I could bring you home.

"I am sorry that you are denied so many comforts, I feel as though it was almost wicked for me to have everything so much better than you can, but I hope you will not think I could not get along without them as well as you, so if you want me to come out to you you must send for me."

Later the same month she added,

"I feel very lonely tonight. I have been alone with the children all day, and the little ones are all asleep, and it is very quiet indeed. Seth, can I ever wait a *whole year* (perhaps more) before I can see you? I read your dear letters two or three times a day, and they are such a comfort.

"Jennie sat in her chair a few days ago thinking very deeply. All at once she looked up at me and said 'Papa will never see me a little girl again, will he?' and she has spoken about it several times. Freddie says he wants to be a big boy when Papa comes. He very often when he hears a carriage asks if it is not Papa coming."

In early December Hannah reveals that, though only twenty-four, she has lost all her upper teeth.

"I have not grown so fleshy that you would not know me, but having my teeth out has kept me back," she wrote. "I expect I shall look very young when I get my new ones. I had to pay nine dollars for taking chloroform and having thirteen teeth out, fifty cents a tooth, and the new ones will be twenty-five or thirty dollars. Everything is so high. Flour is fourteen dollars a barrel, corn a dollar and seventy-eight a bushel, and everything else according.

"Seth, I spend so much money I get discouraged but I try to spend only what is necessary, but you will forgive me if I spend more than I ought, for I do not know how now, but I think I shall learn, and I *try very hard* to do as I ought. . . ."

Her respect for her husband stemmed in part, probably, from the fact that he was twenty-one years her senior. Another reason may have been a good-humored but stubborn precision about Blackmer which is manifest in the following anecdote from his last voyage as master of the bark Active:

Struck by a gale off the Alaskan coast "with hail stones as large as robins' eggs," the Active wallowed in shallow water and breakers until the captain gave orders to jettison 150 barrels of oil on deck, so that "on the fifteenth day, we made the Diomede Islands. We had not seen the sun, moon or stars for several days.

"On the 18th day of the gale I said to the mate 'Today is Wednesday.'

" 'No,' he replied, 'it is Tuesday.'

"I then asked the second mate, and found that he had the same impression. I then asked the third and fourth mates, but they all thought it was Tuesday. I then asked the steward and he said he didn't know what day it was.

“On the next morning when between St. Lawrence Island and the coast of Asia the sun and moon were both visible, it occurred to me that I might get the day of the week by getting the distance between them. I got my sextant and measured the distance, took the distance in the Nautical Almanac from the meridian at Greenwich, and compared them.

“The distance of one day would not do for another, wouldn’t compare at all. I found that I was right, established my day of the week, and showed the result of my work to the mate, who acknowledged that he was wrong, and of course I was pleased.”

And one more note, because this book is not primarily about whaling, or even about whaling masters, but---again---about human beings in a unique time in history:

“On our return home from that last voyage we had a number of sick people on board, and we were anxious to get them down to Honolulu in the hospital. I used to go into the forecastle frequently to see them, and one day I saw a Portuguese from the Cape de Verde Islands reading his Bible to a Central American, who couldn’t understand a word that was read to him, but he thought he needed something of the kind and so listened attentively. I was much affected. It was a pathetic sight.”

Captain Blackmer returned from this final trip in 1871, ending a thirty-seven-year career as a whaleman. Shortly after that he and Hannah moved their houseful of children to Hyde Park, Massachusetts, where he lived until 1903 and she until 1913.

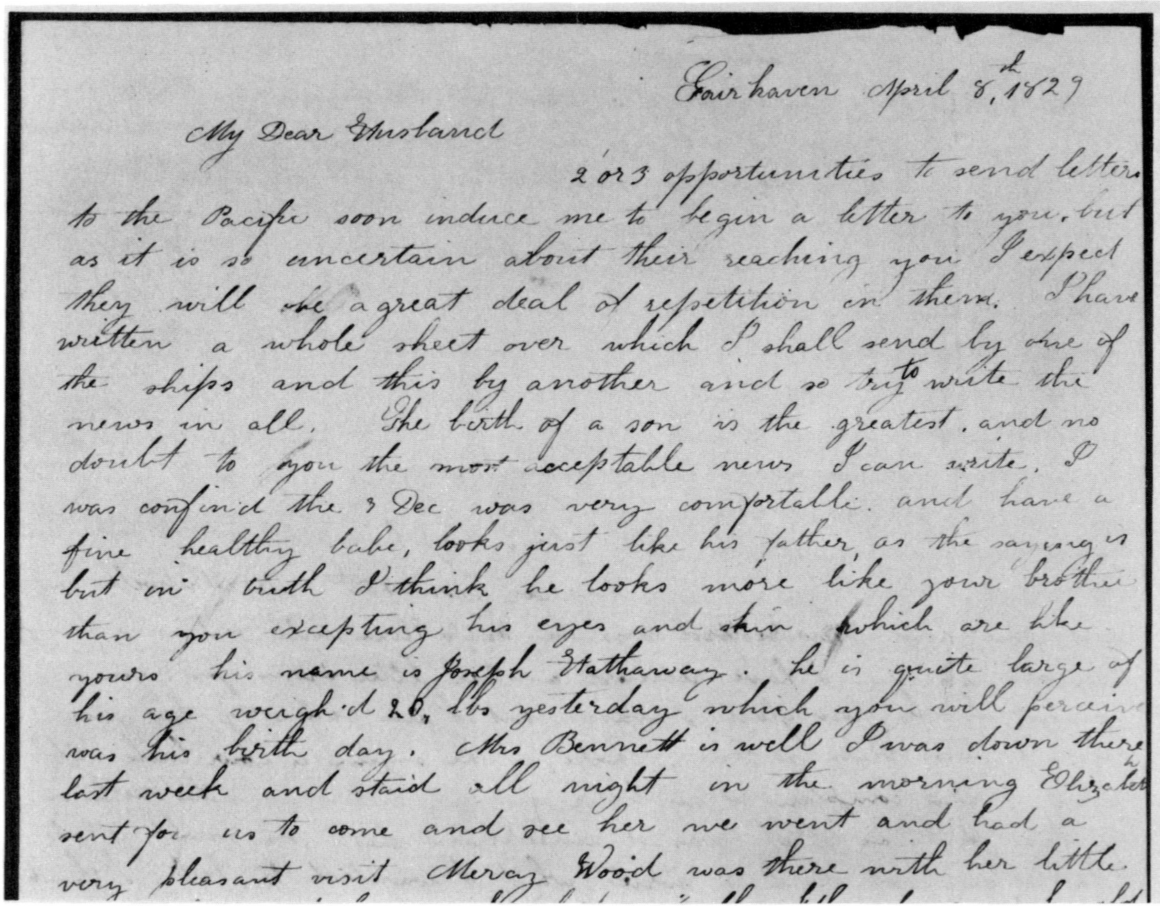

Fairhaven April 8th, 1829

My Dear Husband

2 or 3 opportunities to send letters to the Pacific soon induce me to begin a letter to you, but as it is so uncertain about their reaching you I expect they will be a great deal of repetition in them. I have written a whole sheet over which I shall send by one of the ships and this by another and so try to write the news in all. The birth of a son is the greatest, and no doubt to you the most acceptable news I can write. I was confin'd the 3 Dec was very comfortable. and have a fine healthy babe, looks just like his father, as the saying is but in truth I think he looks more like your brother than you excepting his eyes and skin which are like yours his name is Joseph Hathaway. he is quite large of his age weigh'd 20, lbs yesterday which you will perceive was his birth day. Mrs Bennett is well I was down there last week and staid all night in the morning Elizabeth sent for us to come and see her we went and had a very pleasant visit Mercy Wood was there with her little

Ann Hathaway Burgess wrote the news of their son Joseph's birth to her husband, Paul Cook Burgess, three times before her letter finally overtook the whaler Chelsea, on which he was then first mate, in May of 1829.

ANN H. TO PAUL C. BURGESS

"The birth of a son is the greatest and no doubt to you the most acceptable news I can write. I was confin'd the 7 Dec. was very comfortable and have a fine healthy babe, looks just like his father, Mother says, but in truth I think he looks more like your brother Isaiah than you excepting his eyes and skin which are like yours."

There were eight Hathaways in that Fairhaven clan in all, and if Almira was the frailest and Hannah the stalwart one (who had had the added good sense to marry into the Whitwell money) and Shadrach the most adventurous, sailing again and again "unto who knows where" and settling eventually in, of all places, Indiana, and if each had to have a superlative of some kind, as most men directly or obliquely or at least occasionally insist upon, then Ann appeared to be the soberest. And if as she grew older the sobriety came to mean severe, and the thin downward curve of the right side of her mouth to suggest acerbity and even more, it was perhaps because she was really the one who had known best how to love, though Paul Burgess alone, when she was young anyway, realized it. And the realization was hardly full when it was gone.

"I hear this day that the Canton will sail the 27th and therefore improve this early opportunity to write." She was twenty in that August of 1828, married to Paul Cook Burgess since the January before, and expecting his child in December, though "My health is good for me" was the only reference she made to her pregnancy in that letter or any other. "Mother talks a great deal about you, and John laughs and tells her that she thinks every thing Mr. Burgess says or does is right. She says I must tell you she misses you very much, nobody to dance now you are gone."

Paul was twenty-six, cut a smart and bold-eyed figure, and had made two bachelor voyages around Cape Horn before he married Ann. The seven months on land since their wedding had been the longest stretch ashore he'd known since he was fifteen, but he was off now as mate on the whaler Chelsea out of New London, and Ann had been alone for three weeks. She had one hundred and thirty-one more to go, which fortunately she did not know. So even though there was not much to write from the tininess of Oxford Village, as the

Ann Burgess, widow of Paul Cook Burgess. Captain Burgess was lost at sea on his first voyage as master of the whaler Connecticut, out of New London, in 1831. He was twenty-nine.

northern section of Fairhaven between the bridge and the cemetery was known, she sat at the pine table in the room they had shared in her mother's home, her feet propped over the coals in their fireplace as fall came on, and did so any--way.

". . . The Maria Theresa is in the stream and will sail shortly for the Pacific Ocean, so I resume my pen to you. As it is so uncertain about the letters reaching you I expect there will be a great deal of repetition in them.

"O my dear, I cannot make it you are gone to stay it seems all the time as if you were coming back soon. My Dear Husband what shall I say, I don't want to write I want to talk with you and if I knew you would get this I could say as much again." She wrote a long page then about houses building and funerals and sermon texts, the latter brief, of the health and trials of a dozen friends, and then, "O Paul I have not said any thing for myself and I hardly know what to say. You know something by yourself how I feel, no more at present but this, I thank you for your letters being so long and as you see I profit by your example. Yours in love, Ann Burgess."

She folded the three sheets down from the top and up from the bottom, then from side to side, sealed them with red wax, held her hand over the circle until it stuck firm, then carried the letter next door to her oldest sister Hannah Whitwell, whose husband would take it aboard the Maria Theresa the next morning. The lurch beneath her chest bone when she reached the top of the curving Whitwell stairs was welcome because it was alive, and company, and something she could feel.

Ann Burgess was of course not alone in this sterile pattern, and as the weeks without Paul went on she moved with uneven success into the routine which in some measure preserved the health and stability and quite possibly the sanity of all those New England women caught in a pseudo-romantic era in which so much time was consumed in trying to dispose of it:

"Last week I staid a day and night with Elizabeth Burgess, and Friday last all our family, with Martha, Sarah and Mercy Wood and Philena were invited there to spend the afternoon. Saturday we were all invited to Mr. Goulds and all went but myself. I chose to stay at home, today our folks are gone to Mr. Jenneys. Tomorrow is the funeral of Minister Cobbs daughter in Rochester, she was married a little before we were, had a son which died and she surviv'd it but four weeks. Mr. Tompkins has entirely quit drinking and the temperance society has a great many new members.

"O Paul, if I knew what you wanted to hear I would write it with such pleasure. To say I long to see you is to tell you what you already know. I am extremely sorry you did not fall in with the Eagle as I hop'd to have news. Three times I have dream'd you were dead and distressing dreams they were too. It is nearly sunset and I must say good night. I shall send this by the Richmond, she sails this week. Do not fail of writing every opportunity to your affectionate wife, Ann Burgess."

In the second week of December there was at last something concrete and urgent to tell him, and when Paul after seven weeks had not answered she wrote again in January, and once more in February, and finally on April 8, 1828, a letter which the William and Elisa delivered two months later to Mr. Paul C. Burgess, Ship Chelsea, Pacific Ocean:

"I have written a whole sheet over which I shall send by one of the ships and this by another and so try to write the news in all.

"The birth of a son is the greatest and no doubt to you the most acceptable news I can write. I was confin'd the 7 Dec. was very comfortable, and have a fine healthy babe, looks just like his father, Mother says, but in truth I think he looks more like your brother Isaiah than you excepting his eyes and skin which are like yours. His name is Joseph Hathaway. He is quite large of his age, weigh'd 20 lbs yesterday." And then because there was apparently no ship heading out of New Bedford for the Pacific that week or until midway in the next Ann kept the letter on her writing table and added on April 20, before she was finally able to send it across the bridge to New Bedford and on the first lap of its way: "I expect you will fancy to yourself how Joseph looks, he wears caps now but will not much longer, he is very good when I leave him, Almira took care of him for me to go to meeting yesterday in the forenoon. In the afternoon Mercy Wood was here with her little Lemuel he weighs but 12½ lbs though he is nine weeks older than Joseph.

"O my dear I have receiv'd another letter from you, you cannot think how glad I am but I am very sorry that your passage has been so long, the first part of your voyage so unfortunate, and that you knew nothing of our dear son's birth when you wrote. Aunt Becca says he is clear Burgess. Your affectionate, Ann."

In June of 1829, the rigors of the winter over, Ann took up heart and pen simultaneously by noting that "One third or the time you plan to be absent has nearly expir'd as it will soon be 10 months since you left me. I had hop'd for you a shorter voyage but your precious life has been preserv'd amid a thousand dangers and it must be that you have been upheld by him whose name alone is Jehovah.

"You complain of my letter, he assur'd that the truth is I then felt it was so uncertain that you would get it that I did not dare to write what I wanted to but I think for the future I shall write just as it comes in my head and let it go. Our dear son is well quite fat and hearty. He receives a deal of attention, Brother Paul has sent two frocks and he has a silver spoon from Aunt Read and another from Aunt Whitwell.

"George brew'd yesterday and Elvira came down just now and brought me a bottle of beer, very nice. I wish you could have it. The Herald has got home full and considerable sperm oil. The Midas got in Saturday all full also the Sophia from Cape Horn. Mrs. Bennett is in very good spirits for they have had letters from the Eagle and they had rising 1300 barrels 13th March and would they thought be at home this month."

She put the pen back into the inkstand and watched the sun from the west glint through the window onto its propped end for several minutes before she drew it out again and added in blacker script what she really wanted, and had so often yearned, to say:

"I sometimes think if you go round Cape Horn again I shall go with you for I cannot be left so long if it is possible to avoid it. Yours in greatest love, Ann Burgess."

Clara Nye and William Piatt Burgess were two of the three children of Captain Paul C. Burgess' son Joseph.

It was not until a raw morning in late March of 1831 that Paul Burgess and the Chelsea, the vessel at last full, came back into New London, and that the young father, two days after that, reached home. Joseph was two years and four months old, both introduction and reunion were joyous --- despite the grave illness of Ann's sister, Almira Read, wife of Captain Sheffel Read, three doors down the street -- and the Widow Hathaway wept and then clapped as Paul brought the house alive by dancing, this time with his boy on his shoulder, while Ann sat smiling and trying to sew and awaiting the evening.

Presumably Paul danced again in May --- though discreetly, because Mother Hathaway was deep in mourning for Almira, the first of her children to die -- when he received $1,820.90, one twenty-eighth of the Chelsea's cargo, for his thirty-one months of work and absence. He put away a part for Ann, and a jot toward the later instruction of Joseph (prophetically, Ann thought before a year was out, even knowing he had promised to add to it many times over) and then, despite his brother Isaiah's urging that he abandon the sea for a year and build a house, he put $500 into his pocket and took off again for New London, where on June 9 he bought one-sixteenth of the ship Connecticut, together with "one-sixteenth of her Boat Tackle, Apparel, and Furniture." She was 107 feet long, weighed 397 tons, and was only nine years old, but most important was the fact that Paul Burgess was to sail not as mate but as master.

On June 16 he insured her for $1,000 "lost or not lost, against the perils. . . of the Seas, Fire, Enemies, Pirates, assailing Thieves, Restraints and Detainments of all Kings, Princes, or People, Barratry of the Master, and all other losses and misfortunes which have or shall come to the damage of the said ship and outfits. . . ."

Exactly what form the tragedy took no one is sure. The Connecticut cleared New London late in June, 1831, bound for the Pacific, and Paul did not return. Starbuck's *History of the American Whale Fishery* uses the classic line "died while fast to a whale."

In January Ann gave birth to a second son, whom she called Paul, and who lived until the following July. She waited until December of 1833 before she paid to an Isaac Thompson of Fairhaven $7.73 for inscribing the names of both Pauls on a stone in Riverside Cemetery, and one dollar more for "50 letters extra."

And it was two years after that that the elder Ann, the Widow Hathaway, bought in February the Oxford Village, Fairhaven, house which already knew so much of the fullness and hardness of so many of their lives, and which until then she had rented from Franklin and Mary Ann Bates. It had eight rooms, and almost a half acre of land, and it cost $450, but there were two widows now and it seemed fairly certain that the younger one, who was almost twenty-seven, would not remarry. (She never did.) There was the need, too, to put down a root for her grandson Joseph, who would be seven in December.

BENJAMIN FRANKLIN GIFFORD

"Those who were saved took to the main rigging and went ashore on part of the main deck. . . The Capt., one mate and a boy were found and decently buried, the others where are they?"

Late in 1822 Benjamin Franklin Gifford grew tired of working his parents' farm. He had seldom been away from it, even as far as a classroom. (The first really public school in Fairhaven, a squat stone building on Bread and Cheese Road, would not open until 1828, and few rural families could afford the two dollars a semester charged by the Fairhaven Academy.) There were some big oceans out there beyond the mouth of the Acushnet River; he wanted to see and smell them, and to set his young foot on the lands they washed up on all those thousands of miles away.

By 1825 he had been aboard a New Bedford whaler for more than two years and had begun to feel that the Fairhaven farm, which he was never to see again, had not been such a bad place after all. When spring came he managed passage from the South Seas to England, and in August he sailed from Plymouth, in the brig William Amelia, for St. John, New Brunswick, and home. From here it is best to let five letters — two from him to his father, two from a commissioner of wrecks, and one from the son of a lighthouse keeper on Libby Island, tell what the donor of the letters, who has also provided the bracketed material, calls accurately "the pathetic story."

Letter 1

(To Mr. Paul Gifford in the County of Bristol, Fairhaven, Massachusetts January the 1st. 1823 at Gallapagus (sic) Island)

Deair farther I write to you to inform you that i am well and hope that you air well and i send my love to you all tell Henry not to get marede tel i get back we have struck 7 whales but got none we air now at the gal pagus islands we have got wood and tarpen (terrapin) we air in company with the ship brothers of Nantucket we air well on board now we air gonta start for Japan we haven goat [got] oil yet but i hope that [we] have sum in another 7 months we have sain [seen] the Parnasso [Parnassus?] of New Bedford i see Captain hitch [Hitch] that I was with in Scooner [Schooner] and he was very glad to heair from hom [home] i hope that you have had fine crop this yeair i send my love to you all farther mother brother and sisters you must right [write] when you have noptuty [opportunity] so no more at this time

Benjamin F. Gifford

Letter 2

Note by Benjamin's father: From son Benjamin F. Gifford to Mr. Paul Gifford, Fairhaven May 18, 1825. Letter arrived New Bedford July 2, 1825

Dear fairther i am well and hope you air the same [I] roat to you the other day but that i have now found a frend Lobred [?] Howland sun [Howland's son] he has inform me that you air well and i now right to you by him he tels me that my grandfairther well and i am very glad to heair it he tells me that he saw henry abought fore months a gow [ago] and you were all well then and i was very hapy to heair it from him he has told me all abought home and I was very happy to hair such news from home as i have haird no news sence I had form the ship Progress [?] of Fairhaven i feld [fell] in with this frend sence i rote the other (letter). . . i am in a hurry now i am more ancus [anxious] to come home now than ever so no more at this time

Benjamin F. Gifford

Letter 3

Note by Benjamin's father: the death of our son Benjamin Franklin Gifford sent by Abr. Butterfield Commissioner of Racks (Wrecks) Oct. 1825

Machias Oct 29, 1825

It is my painful duty to inform the friends of the late Benjamin Francis [sic] Gifford that he is not living. I understand that he returned from the South Sea where he had been on a whaling voyage to England in the ship Countess of Myrla. That he sailed from Plymouth in England in the Brig William Amelia bound to St. John New Brunswick in ballast which brig was cast away on one of the Libby Islands at the entrance of Machias Bay on Saturday evening the 15th of this month.

There were 12 persons on board 7 of whom were lost the Captain 2 mates and Mr. Gifford among them. The Captain one mate and an apprentice have been found the others not, nor any of the effects belonging to Mr. Gifford. Some of the survivors thought that he had some English goods within a bundle or chest but none have been found. If they should be found I will render a true account of them to his friends. I have communicated all I know respecting the man. If he is found which is not probable I will see that he is decently buried and let you know. With due sympathy,

Abrm. Butterfield
Commissioner of Wrecks

Letter 4

Machias Dec. 15, 1825

Mr. Paul Gifford
Near the Stone Bridge
Tiverton, R.I.

Mr. Paul and Mrs. Patience Gifford. I yesterday mentioned to Mr. McKellan the son of the keeper of the lighthouse on Libby Island that I had rec'd your letter and requested him to state everything he knows respecting your unfortunate son as he was then on the island and was with the surviving men until they went to St. John and assisted in working with the vessel. You will find his statement on the next page which I doubt not is correct. His father's family is the only one on these islands and have had the case and management of the wreck. . .

It seems the Capt. did not know there was a lighthouse in that place and when they saw it about 5 o'clock P.M. mistook it for the Partridge Island light near St. John and meant to keep off till next morning but a violent gale from the S.E. sprang up in the evening and finding they were going ashore let go an anchor with a chain cable when the brig immediately struck very hard. She soon filled when her decks lifted.

Those who were saved took to the main rigging and went ashore on part of the main deck. The Capt. officers and crew were all forward. There was such a terrible sea breaking that the survivors could neither see nor hear from their distressed friends. The Capt. one mate and a boy were found and decently buried, the others where are they? Something may yet possibly be found— if so you shall know of it. There was a bale of goods on board which probably lifted over the bar between the islands near where the brig struck as it has never been found. The brig struck on the flood tide.

About 3 or 4 weeks after the wreck there came ashore a number of pieces of goods suitable for gown patterns but were entirely spoiled. They were not as strong as paper. I have a few samples about as large as a handkerchief. I have never heard of any money being found except one half dollar among the ballast. I am sorry I cannot give you a more consoling account. . . I hope the next season when the weather is calm something more may be found but it is seldom smooth on the S.E. side of that island.

With sentiments of deep sympathy and the friends of your son I must refer you to that Being who can wipe all tears from our eyes.

Your friend

Abrm. Butterfield

(Enclosure in Letter 4)

Mr. Gifford:

Sir: As Mr. Butterfield has requested some further particulars respecting your son whose melancholy fate he announced in a former letter, I here enjoin the few following as I was informed by the survivors of the crew.

I understand that he had lately returned from a whaling voyage in the Countess of Morley, [sic] that he had received a request from his father to return home and accordingly he embarked in the brig William and Amelia belonging to the same owner, Edward W. Jones, Master, bound to St. John N.B. and after a voyage of 8 weeks on the 15th of October about ten o'clock at night went ashore near Libby's Island light.

I was informed that after the brig struck your son went below and dressed himself in a neat suit of clothes and put a valuable silver watch in his pocket with about 5 pounds in money which he informed the seaman was all he possessed. I did not learn he had any property on board except a valuable chest of clothing none of which was saved except a few shirts which were used by the seamen as they were very destitute.

These are all the particulars which came to my knowledge from repeated enquiries of the seamen who escaped. I cannot conclude without offering you an apology for not writing to you before, the only reason was we were anxiously hoping that the body of the unfortunate young man would be found and decently interred before the distressing intelligence was communicated to you. . . likewise that some article would be preserved that might be sent to the afflicted parents as a token and remembrance of their unfortunate son.

I am dear sir your Obd't humble servant

James McKellan

CHARLES PIERCE TO ELIZA T. PIERCE

"I went to Auckland for repairs from the hurrycane, wrote my owners all particulars, but it seems that some one in Sydney had reported that I was down among the Islands selling everything out of the Ship and would next sell the Ship. . . so they wrote to my agent to hold her till they sent for her, and at the same time sent out another Capt. I had the ship about ready for sea when he came."

Eliza Tobey and Charles Pierce had been married a little more than nineteen weeks when he shipped from New Bedford on October 4, 1859, on what was to become one of the longest whaling voyages ever made out of that port: five years and four months.

"It was the hardest parting for me that morning we sailed I ever had," he wrote to Eliza three weeks out. "It was like drawing teeth, and I suppose you had your share too but never mind, dear, you can have the pleasure of thinking that your husband is first mate in the James Allen, one of the finest that belongs to New Bedford. . . The Capt. cannot bare to loock at his wifes picture yet, I have not loocked at yours but twice."

And then, perhaps an omen of his conviction, two years later, that "I never saw Oil come to a Ship so hard as it comes to this one," he added in that first letter to Acushnet, Massachusetts:

"We worked out that night after the Pilot left us very well, the next day quite good weather, employed in getting ready for whaling, but from that day untill two or three days past we have had one continual gale of wind with the exception of a few hours now and then that we could loock around us to see where we were, in the southern edge of the gulf.

"We spoke the Brigg Mary A. Forest of N. York, dismasted. The Capt. came on board and we let him have a spar and some rigging and I scribbled a few lines to you and would now like to know if it was received. The next day saw a steamer towing a large merchant ship, dismasted, both English, so I think it must have been a heavy gale. good night, my dear, good night.

"Sunday morn, Oct. 30th. I left off somewhere about the steamer towing the ship. In the course of a few days we was running before a heavy gale from NW, at noon hove to thinking we had better do it then than to run much longer. It began to blow harder until 4 p.m. and then struck in a kind of squal and it blowed a perfect hurricane for about eight hours. It roared and howled beyond anything we ever heared before. I did not know but our time had come, several seas boarded us, stove one boat. I stood by the weather rail aft for five hours with the water breaking over me every five minutes. the Capt. felt bad with the rest of us but said little. Supper time came, but I felt no hunger. The water that blew in our eyes came with such force that it made them bloodshoten. Tell Charles to tell Matthews his rigging and spars worked first rate, if not the good Bark James Allen would fared hard."

"I like the Capt. very much so far. He has treated me like an equal and shal lose nothing by it. I would go a long way to serve him. He says he hopes his wife and you will be neighborly while we are gone. He says he thinks you are a *very* fine woman. I tell him I dont think, I know it is so . . . When you get this write once a month to Monganui to wait untill called for, and by ships."

No letters greeted him in the New Zealand bay of Monganui – probably as many messages were lost as delivered in that precarious era—and on March 30, 1860, he wrote:

"The Capt. is on board of the William Gifford, Capt. Baker. He has his Wife with him and so rather than go on shore I chose to talk to you. If I could hear form *you* I should like it very much. darling I am very anxious to hear how you are, if I can only hear you are alive and well."

One of the reasons for his concern about her health probably lay in his own concern about becoming a parent.

"Now my love what are you going to name the little one. I should like very much to see it, if alive, dear me, there it is again. I dont know how it is, wether there is one or not. I pity the man that has to go away and leave his wife with child, for he has many anxious moments, but I hope it all came off right and you will send me the likeness of it by and by and yours with it . . . Do not work to hard, remember you are not very rugged, dont go exsposed to the weather, and if you feel ill have the Doctor, not put it off to long."

We have no letters from Eliza—though Charles tells of receiving a half dozen at a time as the years pass – and there are large gaps, chronologically, in his, but by January 27, 1861, he obviously knew that he had a son:

"I know you will take good care of Henry, for he is my *little* darling. Oh Eliza, it takes you and I dont it dear. Mrs. VanWyke wrote to the Capt. it was the handsomest baby she ever saw, and Maria said the same to me in hers, so I think there is something in it.

"now you must not work to hard scrubbing and dusting. And always have everything brought to the house, never carry things back and forth, that is not a womans place, although there is a great many do it, but i would rather pay for it than have my wife do these things.

"The Capt. received several letters this time. he was very sorry to think he had not written last Season, he had one cross letter from Mrs. VanWyke and I do not blame her, he has written this time and was very anxious I should mention it to you. he showed me his letter to his wife, that is what he had written about me. it spoke very well. I think he bears up remarkable well under the hard luck we have had so far.

"We went on the West Coast of N. Zealand from here, there we had very bad weather, saw a good share of whales, but no chance at them. Later we got one, and came on a gale and blowed very heavy for three days, and saved but 40 barrels. I had a large whale missed, clost to the ship. that was bad. The Second Mate finally struck a whale and we succeeded in killing him but it came on dark very quickly, and still squally. We lost the whale hauling him alongside. it was to rugged to send a boat in the dark to make fast again. would you believe it, dear, I shed tears. . . we felt so poor. We loocked for the whale 4 days but saw nothing of him. We cruised there awhile longer, got three whales, cut them in the night, and lucky we did, for it came on to blow as quick as they were in.

"One of them by Boatstearer struck very well but he is one of those kind, strike one whale and miss two. On the Midle Ground I went alongside of a large whale. he owned himself it was the best chance he ever had. he missed. I can tell you what, I talked some then, you had better believe, it was the last whale he will ever dart at for me . . . .

"The other day Capt. Sargent of the Navy was on board and we were talking about officers spending Money on there voyages, I asked him how much he supposed I had spent on all my voyages. He said 1000 dollars and I told him truly that I had used, put it all together, 50 dollars, to spend in foreign Ports. He was greatly surprised you may believe."

A few days later, though it went to Eliza in the same bulky envelope, he added in quite a different vein:

"I often think how you get along alone and unprotected, not that I doubt you in the least, but I cant help thinking how many virteous women has been led astray by listning to those they supposed there best friends. you know dear, it is the duty of a Husband to guard the reputation of his Wife, for a man knows the artful ways of man better than women. please take this dear as it is ment for you dont know the anxious moments I have thinking of you, and before we came in here and I heard from you I have laid in my berth nights and thought and wondered if you had all the care you required. . . I live in hopes of spending many a happy year with you yet."

Exactly how much time Eliza and Charles Pierce had together is uncertain: perhaps no more than four or five of the seventeen years they were married. The bark James Allen returned to New Bedford at last on February 5, 1865, and there is no way to be sure how long Charles stayed on land. May of 1870, however—by which time he was the father of two small sons— found him north of Melbourne, Australia, possibly aboard the Adeline out of New Bedford, and writing to Eliza of the illness which was to claim his life six years later.

"the cold I took before sailing from home I have not got clear of. When we sailed I toock an addition to it, would cough half the night. I toock my medicine and as we drew out into warmer weather I was some better. I still kept to work, but it was hard, and then as we got into cold weather again, right whaling, getting wet, cold fogs, I grew worse.

". . . the long and short of it is that I have been obliged to leave the ship. I am in that state now that I used to be at home, haying. if I get better soon I shal ship again, if not come home. the ship has gone now and I leave on the 2nd of June down the coast so if there is any prospect for me I shal stop a while longer, and if I do I shal send you some money soon. dear wife I have not forgotten you neither shal I so long as God spares my life.

"The most of my trouble is the shortness of breath but in warmer weather I have hopes it will work off, give yourself no uneasyness about me, dear wife. . . God knows how I would like to come home if I had something to come with. I do not alow myself to spend a cent, I am going to see if I cant save some for you. . . as regards selling the place, you must do as you and Uncle thinks best. Try to get a good price if you do sell."

On June 9 he traveled down to Hobart Town, now the capital of the island of Tasmania off the southern tip of Australia, and called simply Hobart, where he wrote on July 13,

"went to the Consul, he received me very kindly, sent me to place to board and went with me to the Doctors. The Doctor gave me cough medicine and Cod Liver oil now my cough is a great deal better, but some shortness of breath. . . Mr. Colvin the Merchant and a number that I knew here before is very kind, they say I shall not wait long for a berth when I am able to go. so dear wife do not worry about me. I think I am the one to fret, for how you are getting along I dont know. As soon as ever I can I will send you some money, it is in my mind constantly.

"it is now Winter here or I think I should be all right now, although not so cold as our northern Winters, thermometer about 5° above. I keep myself very quiet, doing all I can to get well. I think it better to be here at the exspense of the government than to try and get home with nothing. kiss the Children for me, and love to them, and Henry dear, be a good boy, help mother, and when you go to school try to learn all you can, and write me a good long letter."

As the Winter grew deeper Pierce was advised to go to Sydney "as it is warmer, so I came, my cold begins to loosen, and am much better in hopes to Ship before long. I am greatly troubled thinking of how you are getting along, my greatest anxiety is to get able to earn something to send to you."

A century later Eliza, mentioned in the autobiography of William Crapo as a "scholar of conspicuous excellence, in charge of girls' education" in the old New Bedford High School before her marriage, might have eased those straitened days by resuming her career, but nothing in Charles' letters suggests that she did so.

"Although your lot may be hard, dear wife, which I do not for a moment doubt, I should like to be remembered with the same kind feelings as of old.

"My God I think *my* lot has been rather hard since my arrival home in the J. Allen. No man liveing ever tried harder to get on than I have and I have been going astern ever since but I shal never give up as long as I have strength."

Which is a poignant but perhaps appropriate place to go back, briefly, to those paradoxically happier days when he was mate, and eventually master -- after Captain William D. VanWyke left the ship in 1864 in Callao---of the James Allen.

"Yours of Jan. 5th I received by the William Gifford," he writes to Eliza on July 11, 1862. "She brought it out to me in May. You had better believe I was glad. I never received a letter in my life that done me so much good. And such a fine large Boy.

"I exspect you are haveing very hard times at home, nothing but War, War. Sometimes we get papers with War news in them, 4 or 5 months old. We read & talk on those until we get more but still we are a long way behind the times out here.

"I want when you receive this to get $2,000 dollars insured for me, in case of axident, and in the course of a year perhaps there will be more, but $2,000 will do now. That is about what I have in the Ship. I suppose we will be gone 4 long years, or more, as we are so poor, but never mind darling, 3 years is almost passed."

Two months later a letter to Eliza from Spring Bay, Tasmania, where they had put in for wood and water, reports,

"All well but down spirited, and we hear by the papers dreadful times at home.

"Our latest news of War was the battle at Richmond, the English papers state 25,000 killed on both sides. I hope that none of you are suffering on account of hard times.

"I was allways happy go lucky until this Voyage, nothing troubled me, but now it is different. I find my affections stronger now than the day we were Married. I want when you receive this to write one or two more letters to Monganui and all the rest to Habartown, we shall go to Habartown in about three months from now and next Mar. or April to Monganui, for the last time, so you see I want some letters to both places for Habartown will be our last Port, about a year from now. Write your letters the first part of every month and they will come more regular, for the Mail leaves England about the 20th of every month. . . it seems very hard that I cant get letters but once a year, but that is no fault of yours dear."

A fine though ancient batch was waiting for him when the James Allen arrived in Monganui, ahead of schedule, early in February.

"We arrived here on the 7 inst. and I was made very glad by receiving six good long letters but rather old. the latest date July, but they were very acceptable I can ashure you.

"Oh Darling how I wish I could see you even for a little while. I guess it would take some little *forse* to part us in a hurry, dont you. and little Henry. I do dearly love him and should even if I had not heard he was handsome.

"There was three little Boys on board yesterday and I sat down and had a good chat with them and wondered if Henry would be as large when I returned . . . ."

"Tell Mrs. VanWyke the Capt. is well and received six good long letters from her & give her my respects. tell her I will do all I can to make him write her two or three letters for I do think she deserves them and I often tell him so. I would not have my little Wife feel so bad for any thing. He has read me some parts of her letters, so I know.

"You said in your letters you hoped we could tell something about when we should get home. I am sorry to say I think not much short of two years longer unless we are more fortunate than we have been.

". . . the boat has just come alongside from gamming. One of the men gave me a piece of corral. Now I have three kinds, different from any I ever saw before. I think you will like it much and I have some shells, small ones for frames and some Oak Apples resembling pine cones. very pretty, if they will keep. I gathered them on Van Diemens Land. they call the tree the She Oak, a sort of English Oak. (The shells are now in the home of a great granddaughter in Fairhaven, Massachusetts.)

"You say you are growing old. Do not grow old any faster if you can help it. What I want is for my Wife to loock fair & plump at 45 and 50 and the way to do that is not to make a slave of ones self to young.

"Give my love to Tommy & Johnny. (Younger brothers of Eliza, who were apparently living with her in Charles' absence.) tell them to be good boys and I will do something for them yet. I have had something very near the Kangaroo on board, called the wallaby, but it died. I have several Kangaroo skins to fech home. The Leather makes the best boot I ever saw.

"kiss Henry love for me tonight. tell him Father sent it. How I should like to hugg him real tight. poor little fellow how bad he must have felt when he laid his little face on the carpet and cried, thought he was forsaken but I dont, for I know his Mother loves him to much for that, but I suppose the Weaning time is over now."

In November of 1863 the vessel put in at Sydney and on Christmas Day Charles wrote to Eliza:

"I wish you could look in upon us today. Our Ship has been dressed off with her Flags and decorated with green bushes. she is all tarred, Painted & rigged in good shape. A Capt. of a steamer that came in past us, as we are lying in the stream, said the James Allen was the best looking of any Whale Ship he ever saw in Sydney. that was all *my* work on her spars and rigging, but I suppose you are tired of reading about the old ship by this time, so now for Christmas and holiday days.

"Tomorrow is the long wished for day here. it is what they call Boxing day, very much like our Fourth of July, there is to be about 17 boat races run off, some very valuable prizes to be given to the winners.

"Sydney is located something like Boston, a large City surrounded by large Towns. There is a great many sights to be seen, a splendid public Garden and Domain. Tell Johnny and Henry I saw a Kangaroo there and it made me laugh to se the ungainly thing get about and while it stoped to eat, I saw a rooster jump on its back. It had four legs, the little ones dangling behind.

"Sunday Dec. 27th. yesterday was Boxing Day, bad weather, a fine drizeling rain, but for all that there was quite a turne out. I went on an excursion to a place called Manly Beach down the Bay. Three or four steamers run every half hour. there was all sorts of sport, some quite hard ladies but I shal return as chaste a Husband as when I left you.

"... we are about ready now for sea. we have taken Provisions for about six months and if we think it safe to venture will be home in that time, but dont place to much dependence on us at that time. at any rate my dear Eliza, if I think it safe to come I shal use all my influence to do so, I ashure you, for I am almost crazy thinking of home and cant get there.

"You want me to be shure to say whether you shal get the War risk put on my insurance. yes my love, I think it is a wise plan. I want this about next March, about the time we get around Cape Horn."

On January 3, still in Sydney but planning to sail "the fore part of the week," he repeats the hope of being at home "sometime in June next" and adds, "Tell Uncle Wilson that all the Wars in the World wouldnt keep me 4 years longer, the time has been when I might have staid any length of time but it is different now. there is something there now dearer than life itself that I am constantly thinking of. there may be many more capable of exspressing their feelings for their Wife than I am, but I very much doubt if there is many who loves her more, or will do more to serve her. Eliza, if you thought the day we were Married that I loved you, you can rest ashured that today my affections are ten times stronger. I never loved but one Woman, nor never can, do not fret to think about your old loocks etc, for that makes me love you all the more, for some of that perhaps is my fault.

"When I opened your letter and took out your picture I saw at once you had been sick, and had altered so, it started the tears from my eyes. I never loved you more than at that moment, so cheer up my love, a little while longer, and if it is Gods will we will be with each other again, then we can tell one another our trials & troubles & sympthize with each other... look after little Henrys health, let him play out of doors all he can in good weather.

"I have one specimen of coral that if I can only get it home perfect I think would take the eye of any one. it is blood red like the branch of a tree. One Capt. came on board and saw it, said he would give ten dollars for it, but he had to go away without it. I tell them, all, that they are not for sale, but for my Wife.

"I am drawing to the end of my paper so I must wish you a Merry Christmas and a happy New Year. It was the first Christmas I have spent ashore in 18 years.

"I want you to take things easy, not to fret, not to be to sanguine of my arival in June, for you know there is cruisers about so farewel for the present, from your own true Husband."

If Eliza took Charles' advice "not to be to sanguine" of his returning in June she was wise, for on the twenty-third of that month he wrote from Callao, Peru:

"I suppose you will be rather surprised to hear that we are at Callao, but so it is. We left Sydney for home, got almost to Cape Horn, fell in with large quantities of ice, adverse winds, most all darkness and our men getting sick, likewise the Capt. beatting and banging about to the Southard of Lat 50-00 South for about a month, kept away to the North.

"the Capt. is going to leave here. I am going out with the ship to cruise about 4 months, then to Talcahuano (Chile) and home. I am going to my old ground & I am in hopes to get some Oil. the Second and Third Mates are both going with me and about all the crew. we have hung together so long and are bound to se it ended. I am in very good health and I have been since leaveing Sydney. If I was only shure you were all right at home I should be as happy as these hard times admit. this has been a very discouraging voyage as well as a dreadful long one.

"the Capt. appears to enjoy quite good health in Port, but very poor at sea. he dont know yet whether he will come direct home or not, as times is so hard there. tell Mrs. Van Wyke not to fret, for if he does not come home pretty soon, as quick as he is in a way to make a liveing he will send for her. I have had some very serious talk with him about these things. I brought tears to his eyes and my own too. for my part I am allmost crazy to get home, but as we are down here and it is Winter off Cape Horn, I think it would be better to take a short cruise and then it will be comeing summer. the War I am in hopes will be over and it will be better for all concerned.

"I shall be in Talcahuano in Dec. or Jan. if I find plenty of Whales I shal stay as long as I have an emty cask. I have now about 400 barrels emty casks, seen no whales since leaveing Sydney.

"In this I shall send my Picture from the far famed city of Lima. report says they take the best Pictures of any where. You shal se for yourself. I want you to send Henrys and yours to Talcahuano.

"Since writeing the above I have been to Lima and have had my pictures taken and I have some others for you to put in your Album. I saw some great sights at Lima, it beat anything I ever saw. how I wished you were with me to enjoy it to. I always think of you my love when I am haveing a good time. it is as you say a sort of half enjoyment to be separated so long. it seems like a dream to look back over the past 4 years and think of what has been done.

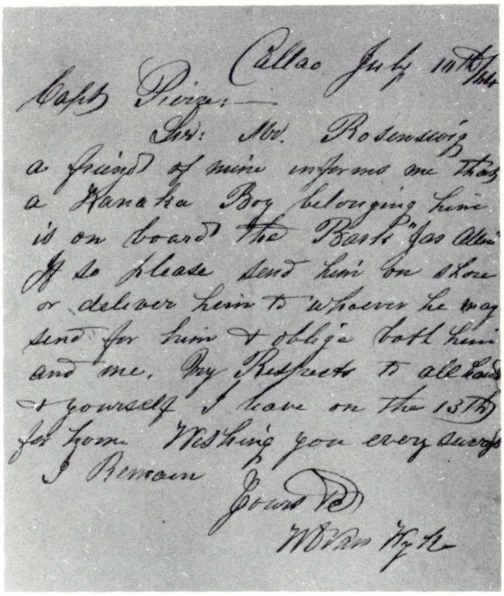

Callao July 10th/64

Capt Pierce—

Sir: Mr. Rosensweig
a friend of mine informs me that
a Kanaka Boy belonging him
is on board the Bark Jas Allen
If so please send him on shore
or deliver him to whoever he may
send for him & oblige both him
and me. My Respects to all hands
& yourself I leave on the 13th
for home Wishing you every success
I Remain

Yours &c

W D Van Wyke

This letter requesting the return of a "Kanaka Boy" was sent to Captain Charles Pierce of the whaler James Allen shortly after Pierce took command of the vessel from Captain William D. VanWyke, who had left the ship in Callao, Peru, because of illness.

I know it must be very hard times there and it troubles me not a little to think of it." Meticulous details about a draft to her for $375, a considerable sum in 1864, follow, and then: "I want my love you should have all you need to make you comfortable and happy. dont stint yourself in the least and then you will please me.

"I know you will feel very bad when you receive this to think we are not comeing direct home. . . after being mate of the Ship for 4½ years, all but one has shipped with me again for me to go as Master. I tell you these things that you may be easy in your mind for it is to bad I know to leave a Wife so long.

". . . I saw Mr. Allen in the Josephine, off the North end of New Zealand, after we came out of Sydney. He told me he saw you and Henry and that the Boy looks very much like his Father. We had a good gam. I can tell you, we talked as fast as our tongues could fly. I had so much to ask him. you do not know how I enjoy reading your letters, I read them over and over, dozens of times. now, my love, believe me the same loveing husband that I was the day we parted."

When the James Allen finally sailed up the Acushnet River into New Bedford eight months later she had aboard, according to the records of Alexander Starbuck, 666 barrels of sperm oil and 454 of whale oil. Maybe it's better not to think about how many gallons a year this catch represented, or how many dollars a man her officers and crew collected in exchange for sixty-four months away from home.

Probably Charles Pierce stayed in Acushnet with Eliza until sometime in 1868 or 1869; we cannot be sure, any more than we can know what he did for a living while he was on land. Whaling seems to have been the only job he knew and loved, and shipping as master the only berth he really wanted, so it is not surprising to find him captain of the whaler Adventurer, out of the port of Sydney, late in 1870.

"I have recovered my health and have got a ship to go whaling as Master, R. Towns and Co Agents, she is called the Adventurer," he wrote home on the last day of that year.

Little did he know when he stepped aboard the Adventurer that he was setting in motion the most memorable events of his professional life -- events which began in disaster but ended, except for his reduced health, in exoneration and a fame which apparently spread from Auckland to Sydney.

"I expect to get away in about a month, fit for about 16 months," he wrote in that December 31st letter. "I get the 72d lay, now what I want is for you to go to Uncle Wilson, or to your Uncle George, for money to live on until I return from this Voyage, then I will send them a Bill of Exchange for their money.

"if only I can make a Voyage then I shal send for you to come to me. believe me, dear Wife, I have not forgotten you. I want you to do as regards the Place as you think best. If I remain out here and you come to me I suppose then you will have to sell. I know of no place that, if I was done going to sea, I should rather live at."

Almost a year later, in December of 1871--though there must have been letters between --he tells, from Norfolk Island, of having 100 barrels of sperm oil, and "the rest of the Voyage proved a failure, not only for me but all the Ships I saw. I dont expect to go in until next Sept. unless I do very well the next six months which I hope and pray I may." Something of the youthful spirit of the James Allen days tinges the rest of this letter, however, perhaps because "I have grate good health since I sailed from Sydney.

"I came in here, found Capt. Macy of Bark Janet he lives handy to Mr. Coes, knows Emily Macy that was. we had a good laugh. . . there is 5 ships of us off the Island.

"By the way you thought my Picture rather over painted, did you, upon my word you did not know you had So good looking a chap, did you.

"dont think because I am sailing out of Sydney that I have forgotten that I am a Husband and Father. I am in hopes to be able to send you some money from this Voyage if I cant send for you. I think you would take lots of comfort out here, so many Captns wifes that you know, and here at this Island they are such fine People. Capt. Frank Bates you know is here. I have been to se him and his Lady. Oh if you had only sent your Picture before Willies I should have it now. I had a late one of his you know before I sailed and none of yours. I do want to see how you look again. how Willie has grown, I declare. he looks as big as Henry. tell him he must learn fast, I do so want to see my boys good scholars.

"I know dear Eliza you have a hard time of it but if it lays in my Power you shal have easy times yet."

The easy times were as elusive as the whales, and the lack of them more painful. One year later, in the spring of 1873, Eliza leanred that her husband, after a series of catastrophes which may well have caused the final undermining of his health, had been stripped of his command of the Adventurer.

"I seat myself again to pen a few more lines to let you know I am alive," the letter of December 2, 1872, begins. "I have written over and over again but can get no letters, or next to none. . . in case you have not received mine I will say again that off Norfolk Island, last Feb. I encountered a very heavy hurrycane, tore the Ship all to pieces.

"I went to Auckland for repairs, wrote my owners all particulars, but it seems that some one in Sydney on my voyage had reported that I was down among the Islands selling everything out of the Ship and would next sell the Ship.

"so when they heard I was in Auckland they wrote to my agent to hold her, till they sent for her, and at the same time sent out another Capt. I had the ship about ready for sea when he came.

"The whole town was in an uproar about it, telling me not to give her up, all the officials gave me great credit for the way I had conducted the whole thing, both before I came in and after I got to Port. and then to think that a man had come to supervise me was to bad. I told them I had done nothing wrong but if the owners heard these reports and believed them I could not blame them. I sent my Books to my agents, they was overhauled and everything in the Ship, and found to be very correct.

"Then my crew declared to a man they would not go with anyone but me as master. I read the articles to them, and advised them to go. I had done all I could for the Ship.

"The crew and officers made me a present of a ring and testimonial--I am going to have it framed. The ring cost $12.50, that is more than ever I see done to a Master before by the Crew." (The framed testimonial, reproduced here, now hangs in the home of the East Fairhaven, Massachusetts, home of a grandson of Henry Pierce.)

"Now when this new Capt. got charge he made the Money fly, draft after draft, and at last he went to sea with what few hands he could get to work her to the Islands, where he shipped men, and then *he* began to do what they heard I done done. He has turned everything into money he could, and then left the Ship. the mate is bringing her home.

"They are very sorry what they had done to me, it has cost them in cash $12,000 besides the loss of what Oil I might have got.

"Tell Willie not to forget his Papa. O God, Eliza, the tears blind me from writing."

Captain Pierce, now 43 and too ill to do physical work, was to have only one more command.

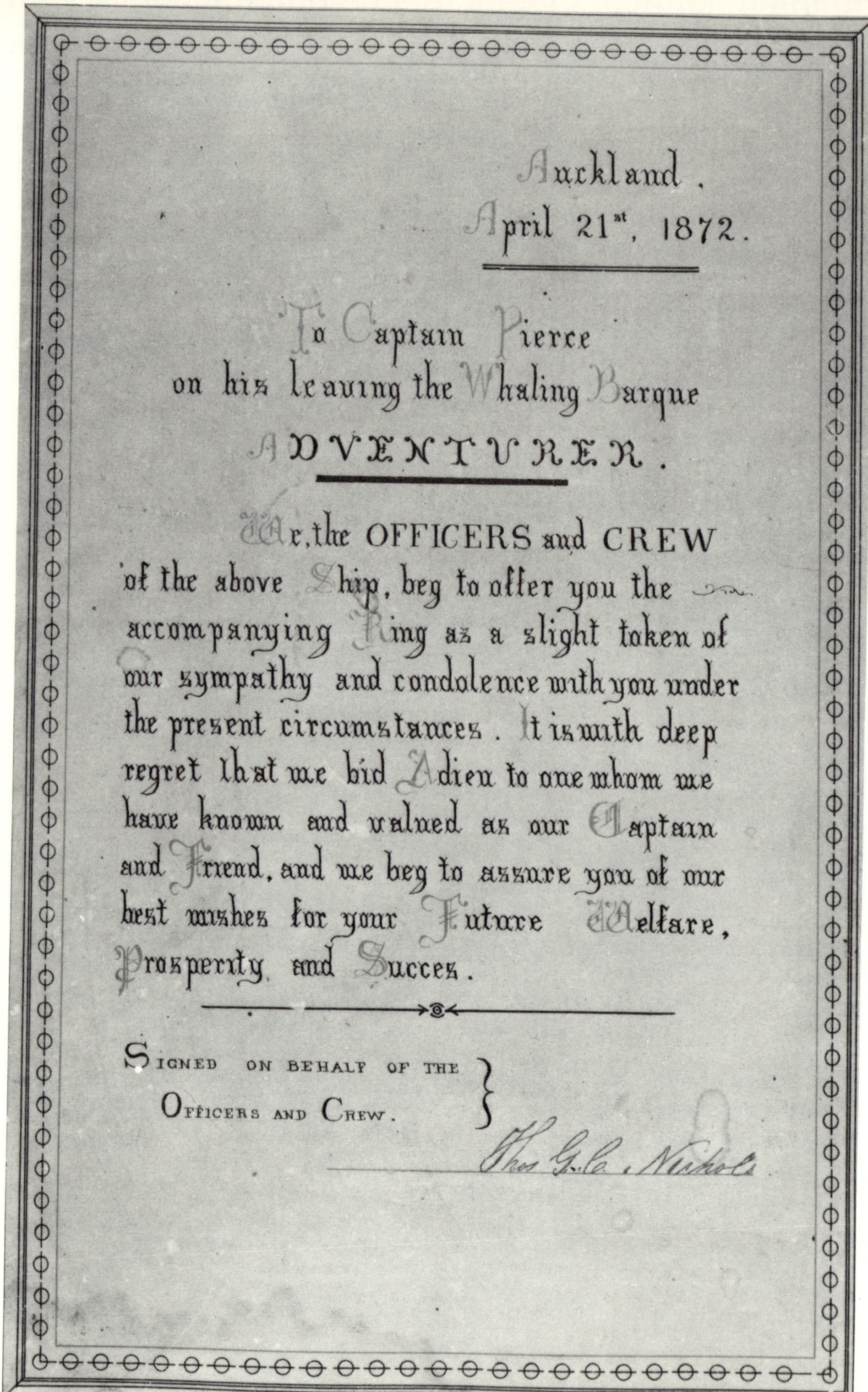

Auckland.
April 21st, 1872.

To Captain Pierce
on his leaving the Whaling Barque
ADVENTURER.

We, the OFFICERS and CREW of the above Ship, beg to offer you the accompanying Ring as a slight token of our sympathy and condolence with you under the present circumstances. It is with deep regret that we bid Adieu to one whom we have known and valued as our Captain and Friend, and we beg to assure you of our best wishes for your Future Welfare, Prosperity and Succes.

Signed on behalf of the Officers and Crew.

Thos G. C. Nichols

This testimonial to Captain Charles Pierce, presented by officers and crew of the bark Adventurer after Pierce had been unjustly stripped of his command in Auckland, New Zealand, in 1872, has been called "one of the most unusual documents in the history of American whaling."

On March 23, 1873, he shipped out of Sydney as master of the bark Chance, even though "I am in very poor health, that is I am not able to work. If I do anything my breath leaves me at once, but if I dont do anything I am all right, have a good appitite and look well except grey.

"the Bronchitis, I think, never will leave me again, but stil I believe that a good Doctor could cure me. I have to be very careful about taking cold, so the owners of the ship told me not to raise my hand to a thing except to navigate the Ship and see that things goes right. and oh if I can only make a voyage for them, and for us all, I shall be happy. it will be my first act, when I get back to Sydney, if I am able, to send you Money."

This letter, which opens "My Dear Wife if Living," repeats the account of the Adventurer disaster, and adds,

"The owners of the Adventurer used me rather hard, and through no fault of mine, as they have since found out. After leaveing the Ship in New Zealand, I went back to Sydney, as I wrote you, and settled with them, satisfactory, and they found their mistake. The reports they heard caused them to feel very anxious, and none to blame either, as I told the people in Auckland when the man came.

"I had the sympathy of the whole city, they says don't give up the Ship. My agent, a very nice man, says What are you going to do, Captain, give her up. I said yes, I dont want their Ship if they dont want me to have her. Well he said it is a shame no man can find any fault with the way you have conducted the affairs.

"I had done more for the port of Auckland, for Whalers, than was ever done before. I made a free Port of it, showed them wherein they stood in their own light. I was the Lion of the day, I had saved the Ship and everybodys life and brought her into Port, and a hard looking Ship she was to, and they gave me great credit for it. There would be a hundred visitors at a time looking at the Ship.

"I toock care and had my Log Book in order because it was an Insurance job, and everything went smooth. All work in these colonys is done by Tender, so the job was tendered for according to my specifications. . . I made a new Ship of her for $5,000 and was about ready for sea when the new master came and a poor thing he was.

"The People in Auckland have formed a Whaleing Company and wanted I should come back there when I had settled in Sydney and they would give me a Ship. So I worked my way back but their ship had not arived and they had to wait for it before buying another. So a friend gave me a passage back to Sydney, there my board was running up and I was almost crazy until the 30th of January last they came for me to go on board the Bark Chance and look after her at $2.50 per day. I was glad I can tell you.

"the agents went to the Owners of the Adventurer and they spoke well of me, then they sent for me to come and ship. I did so, as Master, am now five months out, 30 barrels sperm, 30 Humpback, lying in Humpback Bay. I fitted for one year but you can write to both Sydney and Norfolk, at the same time, then I shal be shure to get one. Now dear Wife accept my best love and wishes and dont think I have forgotten you, or the Children."

Back in Sydney from his last whaling voyage, in mid-March of 1874, Charles rejoiced over long letters from Eliza, and from his sister Elizabeth, but replied in anguish:

"You do not know the hard work it is for me to write home, the misery it makes me to have to write and send you nothing.

"I have done my utmost to get on, but it has been ordained otherwise. If I can get home and among friends possibly I might be able to do something toward yours and the Childrens support but out here I give up, fate seems against me.

"I have been trying to get a passage home, but cannot as I have not the Money, it takes as much as my voyage came to about. I live as economical as I can but still Sydney is a very dear place." He speaks of a "sailing ship, the H. Leichfield, Capt. Spaulding, bound for NY" which might take him, but apparently this hope, like so many others, was unfulfilled, for on July 5, 1874, he wrote from San Francisco,

"I arrived here in the afternoon of the 3rd a very long passage of 78 days, and many hard ships. We went south of New Zealand, very cold out there, I toock cold and was very bad but got better as we drew into warmer weather... to make matters worse our water is half salt, it gave us all the Dysentery. But I have learned not to complain much at what cant be helped.

"I shal stop here a few days to get rested, and recruited, and then try to get East, as they say here. I cant say go home for I have no home, now, unless I go to the Poor House, which I exspect I will have to do. I can see no other way, but again perhaps there may be something I can do to work at something or overlook where it does not require much action. I have no pain, it is my breath. If I work I cannot breathe, so you see there is not much chance of my doing much hand labour. But enough of this."

On July 11, in the last letter we have from his thousands of miles of travel:

"I drop you a few lines to let you know how I am and when I start. I did not dare to go by R. Road thinking it would be much for me, as I talked with quite a number that was lately from the East and said how tired they were, sitting 7 days and nights in the cars, so I thought I would go by Panama, and exspect to start tomorrow the 15th. she is the first Steamer since I have been here. my mind drives me almost mad. people tels me to take things easy but I cannot. Kiss the Boys for me and tell them Father is comeing."

Charles Pierce reached his home in Acushnet in the autumn of 1874, where his illness was diagnosed as tuberculosis or, in the terminology of the late 1800's, consumption. The name for it is unimportant: He was a sick man, and his disease was contagious. So shortly after his return he left once more, but for a far shorter trip to the Brockton, Massachusetts, farm of his sister, Sarah Ann. There he would run no risk of infecting his young sons. And it is from Brockton, on November 10, 1874, that we have the final letter, addressed to his elder boy Henry Pierce, who was then fourteen and a half:

"Now then how do you get along when do you begin your school and do you go to church any. I hope so. I want you should go. A boy never will be much when he grows up to be a man that never went to church when a boy, that is if he had the chance.

Henry Pierce, son of Captain Charles Pierce. From a tintype of about 1874, reproduced -- as are many of the pictures in this volume -- by Nicholas Whitman of the New Bedford Whaling Museum.

"And when you begin high school I hope you will study hard and learn well, get all the Education you can, study nights and remember what you learn, learn it well, and everything you do, do it well and it will work better all through life. There is plenty of time for all those things if we only improve the time.

"Be kind and polite to all and always ready to do a good turn, by doing so it will soon get to be second nature and will gain you many kind friends, for without friends this life is very unhappy.

"My dear Boy, now is the time to prepare for the life of a good man. While you are young, make it a rule to do unto others as you wish them to do unto you, be saving and prudent, dont spend all your small change, in the course of a year it amounts to more than one would think. lay by all you can now to start you in something when you get to be a man.

"Be clean and neat in your habits. pray not get to useing tobacco. you dont know the great amount of money these bad habits takes. think of these things now. now is the time before you enter into these bad habits. it is very easy to keep clear of those things before you get adicted to them, but very hard work to leave them off. I am sick or I would give you a chance to go to school all the time but I am not able to so you must study all the harder for it. You can do allmost anything if you only put your mind to it. I think you are a very good boy now and I want you to continue so and Willie to.

"love your Brother, Henry, be kind to him then he will love you, always be so to each other and you will feel much more happy. go and see your Mother when you can and all your friends. Now I want you should write me a good long letter. Take time and give me all the news. From your Loveing Father Charles H. Pierce to Henry C. Pierce."

A little less than two years later, as the Fall of 1876 approached, Charles Pierce returned to Acushnet. There he died of tuberculosis on October 1, Eliza at his side. He was forty-six years and seven months old.

RUTH B. POST TO FRANCIS M. POST

"there is very much I should like to wright but my mind is so distracted by caire and anxiety that I cannot think of half. . . how I do wish you had sent that ship on a whailing voyage and you staid at home."

". . . Everyone is asleep and it is so still I can hear my heart beat, it seems as if I must hear you speak, as Frank said last sabbath Morning when we was all taking a look at your likeness, he said now let me take it, he looked at it very earnest for a moment and then raised his eyes, Ma, said he, pa is going to speak to me, he has got his lips open now, I will put it to my ear I can hear him whisper if he don't speak loud."

It was early September of 1850 and the bark Eliza Thornton, perhaps the first command of Captain Francis Post which was not a whaler, was laden with cargo and bound for San Francisco. Many New Bedford ships, some of them converted whaling vessels, followed the same path in that decade. The greater part of the lure, historians say, was Gold, but gold or whales -- the effect on families was the same. Especially, perhaps, on wives like Ruth Barker Post, who was left in New Bedford with three children, what seems to have been a slim pocketbook, and the prospect of having to give up her home.

"I expect every day to hear the summons to move out of this house, the house is up at auction, so you must expect to find a house upon the fruit tree lot when you return, and I hope you will return soon not to leave us again."

Captain Post's whaling career had begun twenty-seven years before, when he reached the almost traditional age of fifteen. By the time he was twenty-four he was master of the whaler Huntress, recently married to Ruth Barker, and well into a way of life which was of course renowned for keeping men away from home. As his wife expressed it in the first of those 1850 and 1851 letters:

"Do you ever think of the number of years that has passed away since we was married and not but a few months of that time have we spent together, and yet I do not wish to complain, but I do hope we shall soon be permitted to sit down and enjoy a few hours together from from this wearring anxiety."

The "wearring axiety" may have been rooted in money worries -- Captain Post was on his ninth voyage out of New Bedford and his fortunes, like those of many mariners, rose and ebbed during those risky mid-century years -- in the loneliness endemic among wives of that period, or perhaps in concern for the health of her younger son:

". . . the children have come in they have been up to see Tom Thumb, and such a confusion I have seldom heard, and little Willie oh you should see him oh he is so cute and so good and we all love him dearly, I fear we think so much of him, he is afflicted with that disease about his head as our other little boy was, but the doctor thinks he can work it out of his blood."

The "other little boy," Humphrey, had died of an illness called "brain fever" in January of 1849, when he was a year and a half old, and Ruth was deeply worried about the smaller of her surviving sons, Willie, who was then about fifteen months. It is one of the ironies for which life is famed that it was the older son Frank, the boy who had heard his father's picture "whispering" to him, who died on October 20, 1851, three months before what would have been his sixth birthday. Family records and research give no hint of the cause.

Ruth Post may have sensed that 1851 would be another harsh year for her, for she wrote on New Year's Day to San Francisco, where she hoped the letter would reach her husband:

"I have waited a long time for news, as yet we have not heard a word, we know not what your situation may be at this time, whether you are in a distressed perilous situation, or whether you are as comfortable as you may be at present, God in mercy grant it may be the latter. I have not wrote this two months I have been waiting to hear of your arrival there, but I think I will send this by the first maile, hopeing to hear you have arrived by the next steamer.

"after you had been absent five months I was verry anxious to hear of you there, but since that we heard of the terrible disease, the destroyer of thousands raging there and we rejoiced to think you had not arrived there, truly God is mercifull and orders all things wisely.

"there is much I should like to wright but my mind is so distracted by caire and anxiety that I cannot think of half. . .how I do wish you had sent that ship on a whailing voyage and you staid at home.

"the manless portion of our citizens have been verry much annoyed by a band of robbers that has burrowed in the citty this winter they went to Nancy's the other night and frightened them verry much, he tried the door and windows for twenty moments and Nancy up to the garret window screaming every thing she could think of, at last he started off.

"January the 6. Another steamer arrived and no news for us, we must live on hope and patience two weeks more and then oh then I hope we shall hear of your safe arrival there and if Josiah is anxious to stay. (Josiah Bonney, Captain Post's brother-in-law, had been his companion on many voyages.) You leave the ship and cargo with him to sell and you come home. I know you want to see dear little Willie, if you do not want to see me, so do come home.

he is a pretty lovely little fellow as you can wish to see, you cannot disown him if you want to, for he is the picture of his dady, the children wish you a happy new year.

"we shall move out of this house in three or four weeks and that is no small job."

On January 26 Ruth's spirits rose when "this morning at nine o'clock we received our letters. words cannot express the one half I feel this day. and such a day of rejoicing as we have had for the assurance of your safety and health. . . and yet your letter as it lays before me every word of it is laden with sadness. yet it is verry dear to my heart, oh how I wish I could take dear little Will and fly to you, I could do a thousand things that would render your stay there more comfortable. Do not my dear husband suffer your cairs to weigh to heavy upon your mind. Trust in God and wait patiently and all may yet turn out for the best. I feel to rejoice that you had a pleasant passage, do let me know if your butter and cheese and dried fruit turned out well, and your Cake how did that taste like a chip I suppose.

"oh pa, I wish you could of seen us the morning we received the letters the children danced about as if they was wild and dear little Will as joyous as any of us he stoped in the middle of the room and jumped up and down and spat his little hands and sung out at the top of his voice, pa, pa, pa. aunt Sally came in to rejoice with us and Capt Wilcox and wife at the same time. Capt. Wilcox has been obliged to pay your tax which was $71 rather more than you thought it would be.

"I am looking for a tenement yet and find it verry hard to find such a one as I want Yet if I cannot find a small house I shall take a larger house, I have lived so happily in this little house I find it quite hard to be reconsiled to taking my family into a house with another family of children."

There are no letters at all from February of 1851, possibly because Mrs. Post was looking for a home, finally finding one and, as she reported to her husband from New Bedford on March 8:

"we are all in confusion everything is turned top side down for the want of a label this side up with care, *now dont laugh,* I shall not move again soon as I have taken a house for ten years, I have found a nice little house up to the west end of Allen St. it is rather lonely but I like a quiet place for meditation, it will be verry pleasant and healthy in the summer.

"a Mrs. Toby called here this afternoon to see if you had wrote anything about a box of clothing she had sent out by you to her husband you will please to let me know in your next letter, she is left with four small children.

"now pa, I know you want to hear about dear little Will, he runs all about the house and speaks verry many words plain, I wish you could enjoy his company now he is so good, his health is good now, Francis (Frank) goes to school nicely now and learns quite fast and Lydia gets along verry well, the children wants to see you very much and I do too.

"I hope you have been more successful the last two weeks in disposing of some of your cargo, all though I see by the papers there is verry little difference in the state of the market, I fear for your health the comeing season, you have such a burthen on your mind. Do not suffer your caires to depress your spirits. I see our friends had become reconsiled to a loss on your cargo some time before you arrived there."

Early in April Josiah Bonney, the brother-in-law Mrs. Post had hoped would stay in California to attend to the affairs of the Eliza Thornton, returned alone to New Bedford, and Ruth wrote to her husband:

"I did not think it would give me such unpleasant feelings to see Josiah. I was glad to see him, and yet I could not be reconsiled to see him unaccompanied by you. . . I rejoice to hear you enjoy such good health but it grieves me to learn that it will be almost impossible for you to spend next winter with us, I cannot tell you what a gloom has oppressed my mind the past winter. I feel so desolate, so lonely I almost think I am forsaken. Oh if I could fly to you with my children and find shelter in your bosom it seems to me it would give me more pleasure than all the wealth that can be obtained, forgive me for speaking of my feelings.

"we have moved into our new home and I like it verry well so far, we was fortunate in moveing our things as nothing was injured not in the least. . . Lydia can now go into the forest to get May flowers often this season, and Frank spends his leasure moments sailing boats in the ponds, and bub, dear little pratling bub he speaks all kinds of words and talks from morning until evening, the children are delited with their Dollars. Frank often gets up in the morning tells of dreaming about you.

"Mother Barker has met with a verry sad acident She had a fit and fell from the top to the bottom of the stairs and broke her arms in three places and hurt her head verry bad indeed broke her ribs, and injured her otherwise verry much. She is here with me helpless as an infant.

"I see by your last letter that you have not received but one letter from us dated since your arrival, do not think hard of us as we have sent letters by every mail except two."

In June she wrote that it had been "a verry long month so far as I have not received a letter from you," and added, "I come far short of doeing my duty as I should, come home my love and strengthen me to your strong mind, and I will try to do better.

"your lot looks beautiful, and the trees are looking finely except the peach trees and they are nearly dead, Josiah is as busy as he can be.

"Our City never was in a more flourishing condition than at the present time, Ship building and house building seem to go on as by some magic influence, as some seems to rise others sink down. (Francis Post's interest in the city was strong; he was a public spirited man who served in 1847, from Ward 6, as one of New Bedford's first school committeemen.)

"Capt. Wilcox called yesterday and wished me to say to you that he has received returns from the Gold you sent and it has not turned out well, he requested me to enclose this paper that you may see for yourself."

On September 12 she sent him news of "some verry fine Pears from your lot, and their is some nice quinces but not so many as we had last year, the boys have been troublesome this summer Mother Post has watched them but somehow or other the fruit would walk off.

"Dear little bub cannot realize that he has any father except your likeness, I am in such a hurry to see how he will act, and what he will say, when you do come home. . . he is telling storys from the goose book as fast as he can and he speaks his words verry plain.

"I am now preparing for winter and as I have no help I have to arrange my plans accordingly, the girl who has lived with us so long left us verry suddenly the first of summer, I do not know what ascertained her sudden departure, it may be a prospect of marriage, if so I do not blame her if she can get a *man to stay with her.*

"Capt. Wilcox called to let me have two bank Books I told him I thought it was verry dangerous for such a person as I am to have such a temptation in my possession, he laughed and seemed to think I have not got much spunk I told him I should have if you lived to return for you should never go away to leave any of your family behind you again, and what do you think he said, *I was right.*

"Lydia goes to the bush street school now, and Frank goes to school but he grows so fast he is sick half the time, and now my love, much love from all."

In the same month of September, 1851, after receiving from Captain Post a letter in which *he* had apparently registered a complaint, his wife responded,

". . . you may be sure I was not a little surprised when I read your letter, in the first place to think You should find cause to doubt my affection.

"My dear, dear husband, it grieves me saddly yet I hope you will believe me when I say I have been faithfull in writing to you, and in every other way, if necessary I would give my life to save yours. and yet when I think of your situation, that you do not receive your letters regular I do not blame you in the least if you do scold I will love you the more dearly. . . .

"I steped down Stairs to get tea and Capt. Wilcox called to tell us the unpleasant news that your last package of Gold Dust was robbed of nine hundred dollars on the passage to the mint he has wrote the Express agent in New York and he is going on tomorrow morning, how the affair will turn out we do not know, it is truly unfortunate."

The last of Ruth Barker Post's letters of this series, despite her multiple problems and the most painful one yet to come is a glad one; Captain Post had at last sold his unidentified cargo and had written that he was about "to set sail for home." (Actually the Eliza Thornton burned and he made the trip overland.)

"Oh, what a joyful day it will be when I can hear your voice and your step, and to know that you are with us once more, it seems as if I cannot realize it, it all appears to me like a happy dream, one of those dreams that I have enjoyed and wished they could last forever.

"I did rejoice with exceeding great joy when Josiah read his letter to me. How glad I was to hear you had sold nearly all of your Cargo, and I hope when I get your letter tomorrow I shal find that you are on your way home bye this.

"You perhaps can know the world of anxiety we poor creatures have endured for one year and a half. . . . We are all in good health, what a world of comfort does not those few words express, this is the last letter you will be able to get from us until by the blessings of a merciful God you arrive home."

She could not, of course, have known the cruel inaccuracy of "we are all in good health." Less than two weeks after that October 7th letter, and before his father's return, young Frank fell gravely ill; he died on October 20th of what may have been the congenital condition suggested in his mother's writings.

That Captain Post did not know of the death of his elder boy is apparent in this closing letter from San Francisco:

Barke Eliza Thornton
November, 1851

My dear Son—

Your mother and Lydia Ann have both had letters from me since I have been in California, and as I hear that you are quite a good boy I will write you a little letter which your mother will read to you . . . .

Dont't you remember when I left home little bub was only a baby, and could not even creep, or speak a word. Now he is big enough to run about and speak all kinds of words. I think you must love your little brother very much, he is such a good natured boy.

I don't think the boys out in California have much fun. There is no snow winters to run down hill on, nor ice to skate on. The boys don't play marbles, nor ball, nor sail boats. But then they don't fall into a mill brook as a little boy did I know of. I have seen lots of them sitting in the sand making ovens to bake mud pies in and a dirty set of little fellows they are too. It would do <u>them</u> no harm to fall into a mill brook. But water is not so plenty here. We have to pay 6 cents a pail full for all we use.

We have all kind of people here after gold. There are Americans and Englishmen; Frenchmen and Dutchmen; Chinamen and Kanakas. Maybe when you are big enough you will come out here yourself to see this country and get more gold.

The artist who painted this portrait of Captain Francis Post, master of at least four New Bedford whaling vessels, is unknown.

I am coming home one of these days in the Eliza Thornton and you must come on board of her before she gets up to the wharf. . . Do you think you will know me, I shall be gone so long.

My son, be a good boy, mind your mother, and be kind to Lydia and bub and all your playmates. Then your father will love you very much, and everybody else will love you too.

From your father,

Francis M. Post

Captain Post returned to New Bedford early in 1852, and did not go to sea again. He had made at least nine voyages out of that port: one each on the ship Pacific and the brig Martha; three on the Columbus; one on the Huntress, his first trip as master; one on the Hydaspe, also as master; possible one on the Pleiades, of which he was part owner and listed as captain, and the final voyage West on the Eliza Thornton.

In 1853 he built a house at 122 Washington Street, New Bedford, and in 1855 Mrs. Post bore a son who was given the name of his father, and of the second little boy they had lost: Francis. This Francis lived three score years and a bit more, until 1915. The captain himself died in 1859, at the age of fifty-one.

ALMIRA H. AND SHEFFEL READ

"oh that my dearest friend and earthly treasure may learn the need to keep his lamp forever trimmed and burning."

Almira was probably the frailest of the Fairhaven Hathaway clan, and she was certainly the holiest, but she managed to combine both states of being with a traveler's foot and the compiling of a journal-baedeker which, though it is more concerned with the Eternal than the everyday, tells enough about the life of a childless "whaling widow" to make it pertinent here.

Her husband Captain Sheffel Read was, like most sea-going men of the early 19th Century, absent most of the time. (Starbuck's *History* calls him Sheffield but a family Bible disagrees.) In the late autumn of 1826 he was off on his third voyage as captain of the New Bedford whaler Martha, and Almira, also at sea but bound in a different direction, writes,

"Surely we have abundant cause of gratitude that we were spared through the dangers of the past night. The wind rose to a tremendous gale. In the evening our steerage passengers were very much alarm'd by the waters rushing in through the ceiling, and they made much noise in order to be heard.

"When the hatches were unbarred we ascertained the real cause. The poor creatures, six male and one female, thought there was a hole in the vessel and that we should all go to the bottom shortly. It was a dreay sleepless night, but the wind was fair and under close reefs we made considerable progress on our passage to Savannah."

Fifth of the eight children of Joseph and Ann Dillingham Hathaway, and wife for the past two and a half years of Captain Read, Almira was on her way to the New Englanders' colony in north middle Georgia. In that relatively benign nucleus she would pass the winter with her brothers Paul and Joseph, who had emigrated to Sunbury and Riceborough respectively on the late heels and semi-fortune of their father, who had died of fever there in 1817.

Joseph's widow had returned to the Fairhaven house which had sheltered so many of them and was to surround so many more, expanding and shrinking itself in the process, but she had seen Almira off in 1826 with good grace, hoping that her somber twenty-nine-year-old would find a measure of health and, if possible, something resembling cheer. It was true that Almira had had the summer in Saratoga Springs, where the waters had produced little for her but headache, recurring temperature, swollen limbs, and the distracting proximity of "a Catholick" in the boarding house where she stayed, but let her be off again if she would. She was after all married, and Captain Read had provided the funds.

And so on a blustery November morning the sloop Eliza Nichols, depending solely on the wind for power, cleared New York harbor on the first leg of her nine-day sail to Savannah with Almira and twenty-two other passengers aboard, and a "Captain Blankenship, whose every appearance is of a godly man" in command. Almira's spirits rose and fell with the weather.

"We have had a most tremendous sea, together with high winds and them contrary," she wrote on November 16th. "I have been very seasick. The confined air in being obliged to have the deadlights in adds much to the gloom of our narrow room.

"Much have I thought of my dear, dear husband who for nine months past has been confined on board his ship on this watery world."

On the 18th the ship was favored with a "fine breeze of wind" and the entries brighten accordingly. "This afternoon the celebrated Cape Hatteras was in full view. By kind Providence we have safely passed this dangerous shoal which, at this time of year, we consider peculiarly fortunate.

"I feel quite well, and have been able to read the Holy Scriptures, as well as a number of interesting regligious tracts on board. Good order and harmony prevails. Very little profane language, although I am sorry to say there is one appears not to feel that God will not hold him guiltless who taketh his name in vain."

On the 19th came the gale which so terrified the steerage passengers, but by the 20th the sloop was off Charleston "with a fair wind and delightful weather," and on the 23rd, after an exasperating night of being grounded nine miles up the muddy river, the vessel docked at last in Savannah.

"Parted with several of our passengers, distributed some tracts among the seamen, and left the ship," Almira records briskly of a voyage which, even in a later epoch of bracers and tranquillizers, would have required of many women a recuperation of days. The fact was that Almira, despite an elaborate (and no doubt genuine) bill of complaints, often enjoyed the resilience of a Valkyrie, and that a major root of her physical distress lay in her chronic and Calvinistic jousting with her soul.

"I find every morning cause to mourn that I live at this poor dying rate, so little engaged in the service of the best of masters," she grieves on returning, during that Georgia winter, from a typical Sunday at the Yankee church south of Savannah. The service had lasted all day, with a break for lunch, but Almira had done little more than taste the cold ham and chicken and salads spread out on muslin-covered tables on the grounds; it was more pressing to visit once again the "narrow house" occupied by her father, who had managed to keep shoe-making firms flourishing in their Massachusetts village as well as in Riceborough until he had been buried, nine years before, with dozens of other Northerners around him, in Midway churchyard. "I humbly pray that in future I may redeem time, live like a stranger and pilgrim, and pass the time of my sojourning here in fear."

Far from sojourning in protracted terror, however, she often enjoyed to the fullest, though with the usual painful afterthoughts, the pleasures for which her surroundings were renowned.

"Last night my brothers, sister and myself attended a wedding in Sunbury," she tells in mid-January. "The house was handsomely decorated with flowers and evergreens, and a great variety of cakes and tarts with fruits served up in the greatest profusion. The most splendid supper table I ever saw. Every variety of flesh and fowl that could be produced, and a rich dessert of sweetmeats and jellys, syllabubs, floating island, etc., etc., etc."

The vanished practise of "calling" was an established and active custom in 1826, and almost every temperate day found Almira climbing into her brother's gig and covering an impressive itinerary.

"The ladies in Sunbury are very sociable, but they differ in the time they pass with you, as they never calculate to stay long even with their most particular friends," she comments. "Consequently our visits in return are short."

But predominantly she worried over the life she was sure succeeds the temporal one, and the way in which she and her husband and her sister Ann -- the latter two seemed far from "sufficiently converted" to Almira---would fare in it.

"O that he whom I so tenderly love may share in the awakening influences," she wrote as winter passed and the borough entered the burgeoning month of March. And then buttressing herself against the soft air and "jessamines in greatest profusion" she dipped her pen and added, "o that my dearest friend and earthly treasure may learn the need to keep his lamp forever trimmed and burning."

Six weeks later, on a humid April morning of 1827, Joseph trundled Almira and her sister Hannah Whitwell back to Savannah where they boarded, on the next noon, the sloop Rising States for home. Less than a day out Almira records the "horrors" which doubtless had something to do with the decision of so many 19th Century Northerners to remain in Georgia:

"Captn Pike on Tuesday 17 deem'd it best to go into the Gulf Stream where we were almost shook to pieces. I never was half as sick before and sister and Ann Elisabeth extremely ill. On Wednesday 18 he advised us to be carried into the cabin on deck which we were and I began to grow better immediately and on Thursday 19 began to be able to wait on sister a little as she stood much in need of someone. Friday 20 she was so much better as to be able to sit up and have her bed made and Ann Elisabeth is gaining nicely and we all begin to eat heartily and have every thing as nice as we could expect at sea."

Ashore in Providence once more Almira sprinted out on a round of shopping "and to the post office where I receiv'd a letter from my dear sister Ann.

She writes that God is pouring out his spirit in a wonderful manner in Fairhaven and that she humbly trusts she has now an interest in the Redeemer. (Ann had by this time a nearly-vested interest in the lively whaleman Paul Burgess too, but that story has been told.) Have engaged our passage on the stage tomorrow and will leave here at 8 o'clock."

The last lap of the journey to Fairhaven was forty miles and consumed eight hours, but once arrived Almira "found my dear mother and friends well and Ann rejoicing in the truth. What would I give if my beloved husband was here, but I would not forget the loving-kindness of God in permitting me this day to hear he was alive and well in January and that the leak in his ship, about which I have felt so much anxiety, is nothing alarming."

The next weeks were spent waiting for news of incoming vessels, taking on the "sacred and pleasant duty" of a Sabbath class, sewing, wandering around the family house empty now except for her mother and Ann and herself until finally, on a Sunday morning in late May ". . my husband arrived in an unexpected moment. It is a blessing for which I can never be thankful enough. In the afternoon we attended publick worship."

And that is all, after his voyage of one year, three months and four days, she says of his return. He apparently stayed at home all that summer of 1827, for on August 21st she wrote,

"Mr. Read and myself left for Boston. . ." They spent a night in Rochester enroute, another in Middleboro, and then "proceeded to the capital—think the country beautiful and the city very handsome. Many very elegant buildings, particularly about the market, and common, from the statehouse we had a fine view of the surrounding country as well as the whole city.

"Friday visited Bunkers hill where a splendid monument is erecting, the state prison, and Navy yard.

"There were 290 convicts in the prison. Many apparently in good health, and in the prime of life, some very young and a number of greyheaded prisoners. It was a melancholly sight. the prison is very spacious and neatness and good order reigned.

"The seventy-four on board of which we went surpass'd anything of vessel kind I ever conceived of. 210 on deck, 49 wide, 38 deep. It certainly is a huge building, hope it may never be needed."

The return journey was leisurely, but after that the entries grow sparse until finally, on "Sabbath Eve, August 23, 1828" -- by which time Captain Read was enroute to Patagonia as master of the whaler Albion---she says: "This afternoon for the first time in seven months I attended publick worship." A conscientious distillation of the sermon follows, and the rest of the pages are blank. Almira Hathaway Read, born in August, 1797, died on April 1, 1831, four months before her thirty-fourth birthday.

"Surely the most pious and gentle of women," John Hathaway wrote to their brother Shadrach, who had moved to Indiana and knew nothing of her death until it was two weeks past. And if her sister Ann, so soon to become the widow of Captain Paul Burgess once spoke of Almira as "artless, guileless, and the same old sixpence," so much for balance. Ann had spirit, and even a capacity for temper. It's possible that she resented it now and then when Almira assumed such proprietary responsibility for her soul.

. · . MARRIAGE MAXIMS . · .

The following "Marriage Maxims" are worthy of more than a hasty reading. Gentlemen need not pass them by, for they are designed for wives; and ladies should not despise them, for they are addressed to husbands.

The very nearest approach to domestic felicity on earth is in the mutual cultivation of an absolute unselfishness.

Never talk at one another either alone or in company.

Never both manifest anger at one time.

Never speak loud to one another unless the house is on fire.

Never reflect on a past action, which was done with a good motive and the best judgment at the time.

Let each strive to yield oftenest to the wishes of the other.

Let self-abnegation be the daily aim and effort of each.

Never find fault, unless it is perfectly certain that a fault has been committed, and always speak lovingly.

Never taunt with a past mistake. Let by-gones be by-gones.

Never make a remark at the expense of the other, it is meanness.

Neglect the whole world besides rather than one another.

Never allow a request to be repeated.

Never part for a day without loving words to think of during absence.

Never meet without a loving welcome.

Never let the sun go down upon any anger or grievance.

Never consider any fault you have committed settled, until you have frankly confessed it and asked forgiveness.

Never forget the happy hours of early love.

Never sigh over what might have been, but try to make the best of what is.

Never forget that marriage is ordained of God, and that His blessing alone can make it what it should be.

Never be contented till you know you are both walking in the narrow way.

Never let your hopes rest upon anything this side of the eternal home.

—Selected by Capt. J. H. Sherman when in Australia, 1870.

. . . ADVICE TO MARRIED COUPLES . . .

Draw out the good, the best in each other, by cheerfulness, kindness, and love.

The best house-plant to cultivate is Love. It should be planted in the soil of good and honest hearts, fertilized with consideration and helpfullness, and watered with the dew of kindly tender sympathy. You must not expect infallibility in each other, as no one on earth possesses it. So, wisely use honest praise wherever you can.

Be sure to encourage each other in well-doing.

Try to avoid detracting from the enjoyment of each other.

Honor God in your home that His blessing may be there.

We read in Prov., "The curse of the Lord is in the house of the wicked, but He blesseth the habitation of the just."

It has been said: "Who does the best his circumstances allows, does well, acts nobly; angels could do no better."

It has been said: "Better have a soiled dress than a ruffled temper."

Be careful to preserve your health. Of a certain class of priests, perhaps of the Greek church, I've read they can be married but once. So you can see the force of the saying, "As happy as a priest's wife." His careful conduct preserved her life. If this were the universal rule, I believe some wives would live longer and not be so overtaxed with hard work as to make life a burden rather than a pleasure.

Take fine care of each other and heed each other's good advice.

PHEBE A. G. SHERMAN.

WORLD'S CRISIS JOB PRINT, BOSTON.

Captain James Henry Sherman selected these "Marriage Maxims" in 1870, and his wife, Phebe Gifford Sherman, later drew up the "Advice to Married Couples."

PHEBE G. AND JAMES H. SHERMAN

"One week ago today you went away. Last night I shortened the four years by calculation like this—2 years, 8 months only for active life if in bed by 10 and up at 6. When asleep I shall not realize your absence."

"Went in a sleigh for flowers. Was married to James H. Sherman at Cousin Lizzie's near 7 o'clock eve. 52 persons present. By M. Batchelor."

It was New Year's Day, 1873, when Phebe A. Gifford of Westport and James Henry Sherman entered into the union that was to last forty years, involve two whaling voyages to the South Atlantic and Indian Oceans for her – one with their two small sons---the usual stretches of separation and loneliness, and an exemplum, even for the 19th Century's comparatively stable time, of the solidity of the wedding contract.

"I knew Phebe Sherman," a Descendants' member recalled as recently as 1977. "She was blind by that time. (The sudden unexplained blindness which lasted until her death in 1913 began in 1896.) A very sweet gentle woman and her husband was quite fiery and outspoken. . . She went whaling with him and took the two boys Wilbur and Elijah."

Luckily enough, substantial excerpts exist from the journal which Mrs. Sherman, then thirty-one, kept during the first welve months of her marriage. A succinct stylist, with a gift for undertone, she records in the early pages that "JHS proves to be a Christian, kind, gentle, affectionate husband. He has been blessed with perfect peace." Inevitable the entries which follow cover a range both broad and spotty for they were intended, especially after James left for the sea in October, to serve as a mirror of her life for him.

On February 5, for example, she tells of a meeting with "James' father for the first time" (an ellipsis not otherwise explained) and on February 9, "A dog bit my arm from my patting his head. Went to eve meeting. . . I spoke. God helped me."

On October 8 of that year the whaler Milton cleared New Bedford with James Sherman aboard and Phebe's journal became, as planned, a day-to-day letter to him.

"Today the Milton sailed," she wrote a few hours after he left. "My dear James Henry has gone. I brought his large picture home from Fred Andrews' gallery. After J.H. went I prayed and the thought 'trust' came to my mind. . .

Paid \$7¼ for pictures and frames." The next day she added, "I felt I wanted to draw near to your mother after you left. . . I want to be a comfort to my parents and yours. Mother is kind. One hour today piano exercise."

And on October 10,

"I threw leaves into a brook to see them sail down stream. I take more notice now of things pertaining to the sea. I picked a bunch of leaves for their beauty -- thought I might send them to you but had to throw it away.. They thought it was dogwood. I love you J.H. My love for you makes me desire to do as you wish to have me do though you are absent."

The following morning "mother brought in a black kitten. She says it is a sign of good luck to have a black cat come to us. Mother has slept with me since you left 3 nights. . . I hope you will like this my way of writing daily. It gives you an idea of my everyday life. One hour at piano."

An extraordinarily devout woman, she nonetheless had a keen eye and ear for the human scene. The next Sunday, for instance, on the way to church, she remarks that "Sr. Landers and her little boy rode over to New Bedford with us. She eats parched oatmeal dry and thinks it gives her strength. She thinks, too, it benefits the mind. . . We met Ellen going to Charlie Soule's funeral. Thank God that we are not parted by death. May God help us to live as to be *sure* to meet again.

"I shall carry down our Marriage Certificate, Marriage Maxims and a picture to have framed tomorrow. Father sits in the great chair nodding, the clock has just struck ten. I must go to bed without thee, so good night."

In mid-October Phebe "received a letter from Father Sherman today. He addressed me Respected Daughter and said 'May you always be happy.' William Gifford rode with Father and I from the drinking fountain corner County and Smith down street. He says they battle him about your going to sea. He says he tells them that you acted your faith. . . I have your collar boxes on my bureau and father Sherman's letters to you so placed that I can read your name.

"One week ago today you went away. Last night I shortened the four years by calculation like this--2 years, 8 months only for active life if in bed by 10 and up at 6. When asleep I shall not realize your absence."

Exactly two weeks after the Milton sailed Phebe put together all the journel entries to date and "sent my first letter to you. . . The wife of the mate who is with James Stanton writes to her husband and sends a letter every fortnight. When they went for letters he, the mate, got 4 letters and no one else got any. I want to do as well by my husband as anyone else does by hers.

"I have heard so much of suffering and sickness lately it is almost enough to frighten me. I am almost afraid to eat butter or grease lest I might have a cancer. Thank our Heavenly Father that your health and mine is as good as it is.

"Today deposited 30 dollars in the Institution for Savings. . . William Gifford said he never saw father in tears before the time after you had gone."

And so the journal-letter goes. Again: The comments are erratic, but Phebe was not writing for publication. She wanted to share, as well as she could, her whole life with her husband. Perhaps it is best, in order to preserve her intention, to let the rest of it, the November and December entries, run uninterrupted:

Nov. 3, 1873-Monday

I was weighed by Mr. Almy in Mr. Wood's store—112 lbs. Have called into James Smith's. They had today attended Mrs.. Smith's brother's funeral. They rode home with their horses 32 miles in 4½ hours. Father says your honor consists in my appearing well. God bless you, darling. I love you. I am glad that God can bless you more than I.

Nov. 5, 1873-Wednesday

Four weeks today since I saw my dear James Henry. One month of the 48 gone. Four weeks only of the four years gone. . . Our room is now perfumed with thoroughwort. A syrup of that herb is being prepared for father. He took some cold on our Fall River voyage. I am glad that I can think of you. I love you though you are so far away.

Captain and Mrs. James Henry Sherman with their sons Elijah Lloyd, top, and Wilbur Gifford, foreground. The boys accompanied their parents on a whaling voyage aboard the Mermaid, which Captain Sherman called "our home as well as our business," between early 1883 and 1885.

Nov. 7, 1873-Friday

. . . It looks better here than when you went away. The carpeting which is now down in the kitchen improves the looks of the floor. . . Our card pictures are in my album. Uncle Benjamin said he thought I looked composed. 'Tis a pretty good picture of me. I love you.

Nov. 8, 1873

I am now in my bedroom retired for the night though it is not late. . . We cannot now kneel together in our bedroom to pray but you can kneel and pray in your little room in the ship.

November 11, 1873-Tuesday

This A.M. father and I went down on board the bark Morning Star to carry my letter and roll of papers. . . We had one of mother's hens stewed for dinner.

Nov. 14, 1873-Friday

Today I put your suit of clothes and overcoat on line to air. . . I know that if time remains bye and bye four years will have gone.

Nov. 15, 1873-Saturday

George French and his wife were here. Mother has no members of the feathery tribe now; the last was cooked for dinner.

Nov. 17, 1873-Monday

There is but little snow though the most we have had. . . I called into Mr. King's . . . his son was drowned when with Capt. Caswell after whales. The whale had stove the boat and the men were in the water. It was a cold and rough time. Another boat was fast to a whale. The mate asked Capt. C. if he should not cut the line and save the men. Capt. C. said, "Mind the whale and I will save the men." But he was on the ship and the boats were all off and he could not get the ship round.

Nov. 19, 1873

. . . I called into Mrs. Davenport's and she gave me the 25 cents Eliza borrowed before you went away.

Nov. 21, 1873 - Friday

. . . Mahala (Negus), mother and I rode down to the High School to attend the exhibition. . . There was some good music. Hannah Charity said that they had had news from George Allen. He got a whale 4 days out. . . Father is now lying on the lounge and the black kitten has laid herself on him. He looked up at her but did not drive her off.

Nov. 22, 1873

Have been down-street. Your picture was hanging near the door in a square frame.

Nov. 23, 1873-Sunday

Father and mother have not been to meeting today. Father has a boil. I went to meeting A.M. It was not so good as it might have been. . . May God help you to grow in grace. . . may He keep you from backsliding. May you be filled to enter into life eternal. If I don't, I want that you should. Good night, dear James Henry. If you feel at liberty pray for me.

Nov. 28, 1873-Friday

We had our Thanksgiving dinner, a stewed chicken, the next day after Thanksgiving Day.

Nov. 30, 1873

Today closes November. . . This A.M. the 6th mouse has been dispatched. They have all been caught in our store room up garret. Father does not cough as much since taking Kennedy's Medical Discovery.

Dec. 3, 1873-Wednesday

I had 3 teeth filled. Cost 3 dollars.

Dec. 13, 1873-Saturday

Arose before 6. Blacked stove. Rainy.

Dec. 17, 1873-Wednesday

We went to Aunt Lizzie's. She said she had been almost crazy about your going to sea. She thought it the dreadfulest thing that had ever happened in the family. She said to tell you your old Aunt Lizzie thinks more about your bringing home you rather than oil.

Dec. 20, 1873-Saturday

Betsy Jane (the family horse) has gone today. Mr. Wilson bought her. Brother Potter said None-So-Pretties are in bloom out doors in front of their house.

Dec. 21, 1873-Sunday

. . . I have walked more than 6 times the distance from home to our Meeting House today. I asked God to bless you today.

Dec. 25, 1873

Christmas Eve Mrs. Tanner came for me to go with her to the Catholic church. I went.

Dec. 28, 1873-Sunday

We not only know but see now it is winter. Snow witnesses to the fact.

Dec. 29, 1873

Before I arose this A.M. I thought that it was 21 years since when 10 years old I was converted. . . May the spirit of praise to God abide with me throughout the rest of my life. May you too, Dear J. Henry, experience the spirit of praise. Had you been praying especially for me on the 28 or 29 of December?

Dec. 31, 1873-Wednesday

The sun is shining this last day of the year. . . Went to Watch meeting. I wish you, my dear husband, a happy New Year.

Jan 1, 1874-Thursday

This A.M. I read one verse: Isaiah 25:1. Please read it. Evening at lecture. It was not as pleasant as it was one year ago. . . Don't imagine me perfect and find yourself deceived. I am weak and shortcoming, Help me, darling, by your prayers.

Phebe Sherman's journal closes on New Year's Day, 1874, the first anniversary of her marriage.

On October 24, 1876, the Milton returned to New Bedford eleven months earlier than Phebe had dared to begin scanning the harbor for her sails. Whether James Henry spent the next six and a half years on land is uncertain—he made eight voyages on a half dozen vessels in all--but in any event he fathered two boys, and bought the whaler Mermaid, with the help of his father-in-law Elijah Gifford, in the early 1880's. Labeling her "not a new vessel but a very good one," he records that she "stood us in $45,000 -- ship and outfit -- before we ever set a sail on her," and that "the Mermaid was to be our home as well as our business."

Accordingly the family, Phebe and James and their young sons Wilbur and Elijah, left port in 1883 for a voyage which was to last two years. Possibly the most memorable day of that period, in human terms anyway, came in St. Helena, when on their route from the Cape of Good Hope, Phebe became that isolated but busy port's first woman preacher from the whaling fleet.

"My wife, the boys and I went up to the Baptist Church for morning services," an account written by the captain tells. Apparently the congregation arrived, but no minister. The superintendent of the Sunday School, an earnest but beleaguered young man named Jameson, "went through the preliminaries and a chapter from the Book of Acts." Nonplussed about where to go from there he spotted Captain Sherman, whom he knew had spoken in public, "came down the aisle, and asked me if I would make an address." James Henry protested that he had nothing to talk about "but there was my wife. . . from the time she was twelve her father kept her out of school to study the Bible. She had a knowledge of the Scripture that very few people possess. And she could talk about it by the hour.

"So she followed Mr. Jameson up to the front of the church to take her place on the platform. As she walked she silently asked the Lord to help her. Mr. Jameson read from the third chapter of Acts, and she took the same for her text.

"Well sir, it was something great. And a surprise to everyone. There was an audience made up partly of ships' crews but mostly from the regular congregation of the Baptist faith who, we afterwards learned, were opposed to women speaking in public. It was their idea that the pulpit was no place for the female sex.

"The minister never did arrive that morning so the audience had to listen to a dose of woman preaching. That was a strange meeting and I'll never forget it. The remarkable thing was that through force of circumstances they had to listen to my wife expound the Gospel. I said afterwards that I should have been glad if the President and his cabinet had been there to hear her."[1]

In 1888 Mrs. Sherman traveled alone to the Indian Ocean, by way of Europe, to board once more the Mermaid, then on her third voyage under Captain Sherman. Never in very robust health, the doughty little woman nevertheless made the journey without serious incident, and with $300 sewed in the hem of her dress. (She also carried a "to whom it may concern" letter from an H.L. Hastings of Boston, editor of a periodical called The Christian, asking that she be given "advice or assistance regarding routes, steamers, etc.," along the way.)

She remained with her husband until May 13, 1889; less than twenty-four hours after she sailed for home Captain Sherman began, from St. Helena, a five-installment letter to her which he finished fourteen days later:

"Dear loving Phebe, my precious wife to you thus early I begin to write... When I went to my room after supper I shed some tears, *You my Darling was gone.* I remembered that you said the thought 'hope' came to your mind and I was glad of that. After a while I went to bed and slept under three o'clock, after that didn't sleep but little. We have been stowing down oil today....

"By the papers that came today at dinner time it seemed to me that you had been gone a long time when it was in fact only about twenty hours. May God bless you and take care of you, my dear... Good night now, it is ten minutes to nine o'clock. I shall retire soon. I feel somewhat in a hurry to go to sea so as to fill our casks sooner if it be the Lord's will."

On May 19, 1889, the letter begins an account which, though sketchy, reveals much about a Yankee skipper's attitude toward miscegenation or, more specifically, toward a young crewman on the Mermaid who had fallen in love with a native of St. Helena:

---

[1]From *Whaling Voyages of Capt. James Henry Sherman,* p. 156, pp. 162-164.

"Yesterday morning I decided not to discharge Nelson to remain on the island, for you know just as well as I the spirit of his father's letters.

"They his parents are bitterly opposed to his marriage with a colored girl. no matter how good. I told him his mother had rather see him in the cemetery than allied to a colored girl. I told him before the Consul I was acting in his father's place. I would not want my son left on this island for four months under like circumstances.

"The Consul told him that an 'idle brain was the devil's workshop.' He begged to be discharged, but it was no use, my mind was made up. . . The Consul talked very plain to Nelson. The next plan now is to get Nelson a chance to go to England in a passenger ship and I have consented to that if he can make arrangements to go. She sails day after tomorrow so we wait further developments. I want to sail next Wednesday. That will be the 22nd."

On the 21st Captain Sherman had discovered that "there seems to be but very little honor in a great many people here and elsewhere. There was no chance for Nelson to work his passage in the passenger ship and the Capt. asked him $40, so that ends the whole thing about his going home until the schooner comes out or we go home.

"The Consul has written to Nelson's father. I will send it you unsealed. You can read it then send it along to him. I think it is to the point. I will send you a copy of an agreement with Nelson which I wrote and gave to him at <u>his</u> request."

The "agreement" has disappeared and so ends the tale, or as much as is available, of an unhappy young man born in what was for him the wrong era.

Back to routine business, and just before leaving St. Helena several days later than he had planned, Captain Sherman began to wind up his letter to Phebe with:

"Earnest Chambers has run away. He wanted his discharge and I said no. That leaves 26 of us. There are plenty to ship but I shall try to get along with what we have, less to feed you know. Have bought a lot of potatoes and pumpkins and some onions. Have bought over $40 worth of canned meats and four sacks of rice, so as to have something to prevent scurvy when the vegetables are used up.

"It takes a good deal of money here, this time it will be $1200 that I draw on you for. Over $160 to Mr. Watson. . . he wants his pay increased. He shipped last year for the 1/40. Now I agree to give him the 1/30 lay, to date back to time of shipment and continue as long as the ship stays out. Mr. Obey has the 1/25 lay and does not agree to stand watch. I see by the books at the Consul's that he had the 1/22 lay and $200 in advance in the last ship he was in.

"I hope you got along nicely going home and were well and enjoyed the passage and your visit to London. Tomorrow is the Queen's birthday. I must finish all my business tomorrow if possible, also this letter. I hope the boys are well and very loving and kind to you, my dear loving wife.

"I hope you will have no trouble in meeting the bills and hope the loving Father in Heaven will help and guide you and in his providence bring us to meet soon under happy satisfying circumstances. I must retire now, without you my dear, so good night — on paper, it is the best I can do."

And one final line, the next day:

"Now my darling I bid you adieu for the present. I send letters with this to the boys. I am in a hurry. Pray still for your husband. Write and send every mail, if only a line. To the ship now.

James H. Sherman"

Most records indicate that Captain Sherman sailed the Mermaid into her home port of New Bedford for the last time in 1890, when he was about forty-six, and that he spent the next twenty-three years on land with Phebe, who died in 1913.

"She had twice been to the Pacific coast, had toured England and Ireland, and made two whaling voyages to the South Atlantic and Indian Oceans," a New Bedford news account reported at the time of her death.

"Once she went to England alone and from there through the Mediterranean and Red Seas to the Seychelles Islands, where she joined her husband on a whaler and came back to this city.

"At various times Mrs. Sherman wrote verses and prose articles for secular and religious papers. In 1896 she became suddenly blind and learned to read books in the Braille print. She was deeply interested in social and religious work."

James Henry Sherman, who was to live another twenty-eight years and to be just twelve days short of ninety-eight when he died in December of 1941, spent much of his time, after he retired from the sea, as owner and manager of the Sherman House, a hotel which stood on the site of the present post office in New Bedford. The land was taken from him, by the fiat known as eminent domain, a year or so before construction of the post office began in 1913.

TRACED FROM 1969 MAP PUBLISHED BY NEW ZEALAND DEPARTMENT OF LANDS AND SURVEY

CAUTION

All offshore islands, rocks, reefs and shoals are not necessarily shown, nor are they necessarily in their correct position.

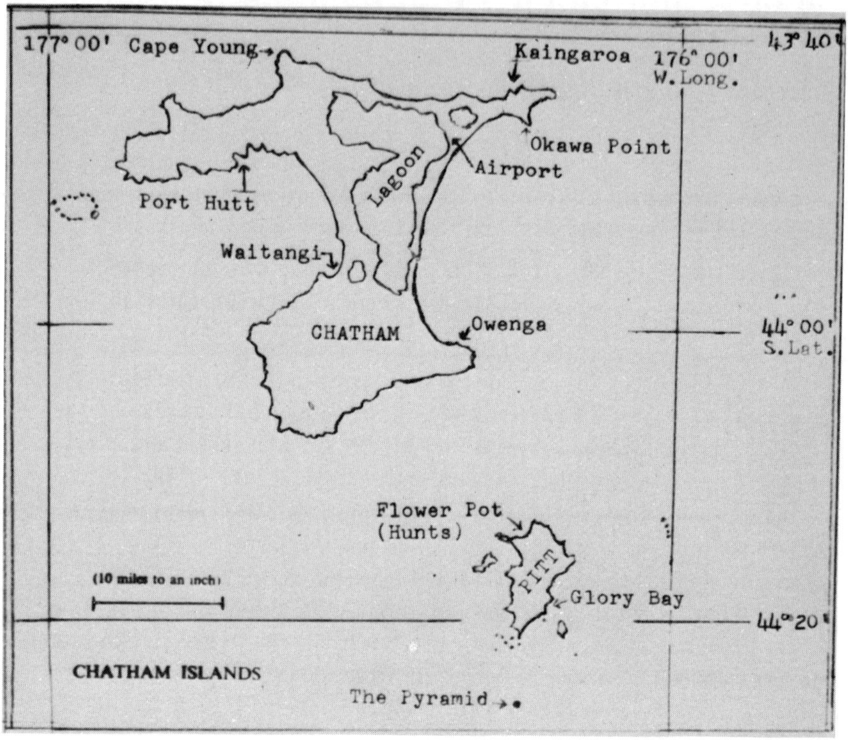

Tiny Pitt Island, five hundred miles off the east coast of New Zealand in the South Pacific, was a popular stopping-place for whalemen in the middle and late 1800's, but has proved almost inaccessible to would-be visitors of the 20th Century.

EMMA AND ALBERT A. THOMAS

"Oh I tell you Mother they make a great time of Christmas here on Pitts Island cooking two or three days beforehand and they are dressed up as though they lived in a city and was going away visiting."

Whaling era children may not have traveled in the style of, say, an Amy Carter, but they went to more remote places and they stayed longer.

Ernest Thomas of New Bedford was six years old when his father, Captain Albert Alexander Thomas of the whaler Merlin, put him and his mother ashore in late 1873 at tiny Pitt Island, 500 miles east of New Zealand, and left them there for two and a half months. The master was not, our somewhat sparse records confirm, tired of them; he had made four voyages without his spirited wife Emma and he liked having her aboard, except that she was forever seasick.

For this reason he made it a practise to leave her on friendly land at intervals, usually Pitt, and to pick up up again, 70 to 100 barrels richer if he'd been lucky, when he was heading for new waters.

"Mrs. Thomas very sick," the log of the Merlin's 1872-1876 voyage records on July 2, 1872. And on October 3, after no reference to her at all in the interim, which tells its own quesy story, "Mrs. Thomas at supper."

The irony about Emma and Ernest Thomas' staying so long at Pitt, which is the smaller of the only two inhabited islands in the South Pacific group called the Chathams, is that today's visitor, despite 20th Century miracles and marvels of transportation, often finds no way to get there. (Sylvia Thomas, granddaughter of Albert and daughter of Ernest, finally landed on nearby Chatham Island in the "passenger capsule" of a World War II Bristol air freighter in 1977-- after Gargantuam effort and labyrinthine planning-- but "rough seas during the full week I was there made it impossible for me to realize a hoped-for visit to Pitt Island.")

Young Ernest, however, put in there several times between his fifth and ninth birthdays, and a high old festival he had of it on Christmas, 1873. He and his mother were guests of the Fred Hunt family, who had left England to settle in the Chathams in 1842, more or less "owned" Pitt Island, and made a substantial living selling provisions to mariners from a cottage near the anchorage called Flower Pot—a cottage built from planks of the New Bedford whaler Franklin after she broke up on Pitt reefs one story April morning in 1859.

"Ernie went a Christmas boxing, as they call it," Emma wrote to Martha Wilcox, her New Bedford stepmother, in January of 1874. "He got up bright and early just as soon as there was light enough to see his way up the hill. It was between three and four o'clock in the morning. All alone by himself he dressed himself and opened the door still so I would not hear him, and when he got up there he went to their window and sung, I wish you a Merry Christmas and a Happy New Year, a box full of money, a barrel of beer, please give me a Christmas box.

"He woke them up singing that and it pleased them very much. Then he got the little boy up there and they went up to Mrs. Langdale's and they sung it under their window and they got 62 cents apiece from there and Ernie got the same from the first place; and then they both came down here and went to Mr. Hunt's door and sung it and he gave them 62 cents apiece and Mrs. Hunt gave them 50 cents apiece and one of the sons gave Ernie 62 cents so you see Ernie made $2.98 cents. I am going to get him something with it so he can keep it and see what he got from friends on this Island."

The "friends on the Island," many of them Hunts or related to Hunts, probably numbered twenty-five or thirty; even today there are only fifty inhabitants on Pitt, several of them descendants of the original British family.

"Christmas we was all together down here excepting Grandfather Hunt and he was too old to get down the hill," Emma writes in 1874. "Then the next day we was all invited to the Daughters that lives to Grandfathers and the next Sunday to the other daughters, Mrs. Langdales. Oh I tell you Mother, they make a great time of Christmas here cooking two or three days beforehand and they are dressed up as though they lived in a city and was going away visiting.

"We had for dinner, Geese, Ducks, Chicken pies, boiled ham pies, plum pudding, and for tea we had a stuffed chive and boiled ham and cake and cheese cakes. They are splendid. . . Ernie has taken solid comfort all the time we have been here."

On that same Christmas Captain Thomas, to whom whaling was strictly a business, devoid of comfort solid or otherwise, is accounted for only in the Merlin's one-sentence log entry: "Begins with moderate west breeze, veering to the SW, blowing a gale at seven AM, wore ship between Star Quay Reef and the Pyramid, heading SSE at midnight, finbacks and porpoises in sight."

Though whaling ships' logs were kept, ostensibly, by mates, they were often written by several hands. In the Merlin's case, indications are strong that the captain took over every now and then and, departing from the rigid third-person style of such documents, made such comments as: "Oh! how I would like to be at home, whaling is so disagreeable," or, "I wish we were going to New Bedford *tonight*." While the mate or mates may well have felt this way, it is unlikely that they would have had the temerity to say so in anything but a private journal. It is probably significant that Captain Thomas, who of course

This cottage near the 19th Century anchorage called Flower Pot, on Pitt Island, was built from planks of the wrecked New Bedford whaler Franklin. The picture was made in 1860, probably by Fred Hunt, an Englishman who bought Pitt from a Maori chief in 1842, became a lifelong friend of Captain and Mrs. Albert A. Thomas, and furnished countless whaling families with fresh meat and vegetables from the little Flower Pot headquarters.

could say anything he wanted to, retired from the sea when he was forty-two and became an orange-grower in Florida.

Part of the reason for his apparent discontent during the Merlin voyage, his fifth and last, may have been the frequent shore leaves of Mrs. Thomas; another could have been his crew, seven of whom deserted in various parts of the South Pacific. (One even jumped overboard when the ship was scarcely out of New Bedford, but was hauled back in, and the cooper was found hiding aboard the whaler Rainbow, in the Bay of Islands, halfway through the voyage.) Four others were discharged, one "disappeared" – the difference between this status and desertion is unclear – one remained on Pitt in February of 1873, and two died, one of dropsy and the other of consumption.

Probably the Merlin officers' punishments of miscreants were no more severe than those employed by many of their colleagues, though there are a half dozen references to irons, one to "pounding irons" for a chronic offender, one case of a crewman tied to the rigging, arms up and taut, for two hours, and another of a whipping with "the fag end of a rope."

Provocation is reflected too, however, in an entry from Roratonga – otherwise a place of delight because of fresh fruit and benign climate – that:

"Charles (Indian) came down to boat intoxicated, after we had been waiting nearly an hour, he being very abusive. Put him in irons, he said he would raise a mutiny, that four or five men could not run the ship, and further he said he missed the whales on purpose." Charles was removed from irons, presumably chastened, the next day; whether he went on intentionally missing whales the log does not say.

But back to Emma's more cheerful correspondence from Pitt to her stepmother, in the first month of 1874:

"When we came ashore Albert brought his hens and one of them has stole a nest and this morning she came off with six little chickens and Ernie said be sure and tell Grandma about it and I have set a hen with 9 eggs and when she comes off and I get back, then I will tell you what kind of luck I have."

The "getting back" refers to the fact that Emma and Ernest were once again going to sea.

"The Merlin came in this morning and Albert is so lonesome without Ernie and me that we are going out with him to stay a little while so I thought I would first drop you a line or two and send by the Mail Schooner for she will be here before we come in again.

"I expect I shall be real seasick for I have been on shore for so long but Albert will take good care of me and I shall get up after awhile. Oh if I could only get over being seasick I should like it."

The prospect of illness did not deter her from anticipating in this letter "the good things they are taking on board for us, new potatoes, Onions, Beets, Cabbage and some live Sheep and one sheep cooked and up in its own fat, jam, cheese cakes, goose berries, eggs, milk," but it is doubtful that she enjoyed many of them. The vessel left Pitt Island on January 25, and on the 27th the log records: "Heavy NE wind at nine AM blowing a gale, took in the foretopsail, at sundown shortened sail."

On February 14th the Chathams were in sight again, which must have been a welcome Valentine greeting to Emma, and the next day the Merlin "came to anchor, Capt. and wife on shore."

Mrs. Thomas may well have stayed on Pitt for another two or three months, during which time the Merlin battled squalls, frequent hail, and leviathan mammals in and around the Chathams. These grounds, according to Sylvia Thomas, "were on the route of the sperm whale migration during the 'Down Under' summer months of December through March," a fact born out by the log's noting on March 12:

"Mr. Joseph raised sperm whales."

To which was added, on Friday the 13th of all days:

"At nine AM lowered 3 boats, at 11 Mr. Nye struck a whale, boatsteerer killed him with iron, took all of Mr. Joseph's line and part of mine, at six PM whale alongside, comenced cutting at sundown." The catch "turned up 71 barrels."

On May 14 Emma returned to the Merlin and remained for the final thirteen months of the voyage.

Dressed as formally as if they were in 19th Century London rather than on an isolated, microscopic island in the South Pacific are Mrs. Albert A. Thomas of New Bedford, standing at right, and four residents of Pitt. The only two identified in this 1873 picture are Ann Hunt Langdale, seated left, and her sister Elizabeth Hunt Gregory, seated center. Both were daughters of the Thomas family's friend Fred Hunt, who emigrated to Pitt from Great Britain in 1842.

". . . brought off from Pitts nine boat loads of wood, about four tons of potatoes, one beeve (large Bullock), 3 sheep, a few beets and turnips," the log of that day tells. "At dark, Capt. and wife aboard."

She went ashore later that month, briefly, at Poor Knights, in the Bay of Islands, enjoyed gamming a number of times with the captain's lady and children on the whaler Atlantic, put her foot on land again at Oparo, where on August 5 the captain traded "for 49 hens about 12 bushels of taro," and then spent four months at sea in the region labeled only "Offshore."

Christmas of 1874 found the Thomas family somewhere around the Marquesas Islands, where Emma must have thought of the Yuletide of a year before, with its dressed-up Hunts and their laden table, and of writing her stepmother: "I often wish we could get to your closet I am afraid we should make sad work in it" — a reference to her hard-pressed and maybe inadequate wardrobe. Our only concrete note on that 1874 Christmas, however, is one succinct line from the Merlin's log: "ENE winds, all sail set, porpoises in sight, very warm."

Back in the Chathams by February of 1875 the Merlin once again anchored at Pitt, "the Capt., wife and boy" went ashore, the vessel spent more long months whaling and finally, in January and February of 1876, approached and rounded Cape Horn (Emma's fate here is not recorded), stopped at Pemambuco in South America—now Recife – crossed the equater on April 4, whaled the Twelve-Forty Grounds and, at last, on May 21, began the joyous part of the cruise called "Homeward Bound."

On May 30 the log-keeper allowed himself to repeat "I wish we were going in to New Bedford tonight," but it was two weeks later before they were "packing up," had "throwed the tryworks junk overboard," stowed the mincing machine and other gear, and on June 19 "spoke a schooner called the Gayhead. . . bore NE about 20 miles, hauled up the cables and got anchors off the bows, Pilot on board at one PM, at ten PM came to anchor off Clarks point."

In elaborate black script about three inches high the log-keeper then wrote END, followed by:

Saw sperm whales 56 times

Saw right whales once

Sent home 681 barrels

Albert, Emma, and young Ernest left for Florida and the orange groves that same year, and "the boy" had little to do with seafaring until the Spanish-American War, when he served as commander of a converted ferryboat for the United States Navy. His life work, however, was land-based; after graduating from Massachusetts Institute of Technology he became Principal Engineering Assistant of the Topographical Bureau of the Borough of Queens, New York, a post which he held more than thirty years.

He died in 1939 having, his daughter Sylvia notes, kept his passion for fruit, which began at Roratonga, throughout his lifetime."

The Merlin, on which Emma and Albert Thomas and their son Ernest cruised the South Pacific in the 1870's, is shown on the ways in this 1880 photo lent by the New Bedford Whaling Museum.

Adelaide Baker Wicks and her husband, Captain Rodolphus Delano Wicks.

ADELAIDE B. AND RODOLPHUS D. WICKS

"One of the men on the island sent the Capt's wife a live chicken and some eggs because she had been sick. The chick is walking around the deck eating coconut."

The first thing Mrs. Rodolphus Delano Wicks did after Captain Wicks opened "the calico box" was to make some turkey-red curtains for their cabin. After that she put together a pink print dress for their daughter Helen, who was two, and cut out a pink sacque for herself.

It was late in 1888, and the master's wife and child had sailed from San Francisco aboard the whaler Coral after crossing the continent from New Bedford by train.

"an accident on the road last night delayed us four hours," Mrs. Wicks wrote on November 16 near Rock Springs, Wyoming. "two freight trains ran into each other and one of them ran into a house belonging to the railroad nearly demolishing it. very pleasant going through Echo moonlight. . . ."

Adelaide Baker Wicks was the captain's second wife, and young Helen his second daughter. His first marriage had ended, as did so many of that era, in the sorrow of returning in 1875 from a four-year voyage to discover that Isabella Baker Wicks had died, at the age of twenty, almost two years before, leaving him a daughter Cora, who must have been about three when she met her father for the first time. Doubtless he would have heard both pieces of news, the joyous followed by the reverse, if the whaler Sea Breeze--his first command as master --had not been hemmed in for three years by Arctic ice.

Ten years went by before he married Isabella's cousin Adelaide on January 13, 1885, and left for San Francisco, the port from which he often began one of his sixteen or more whaling trips. That it was to be a happy union is suggested in a note from him to her on March 13:

"Addie it is morning. I have just got up half past seven. Oh my dear Addie how can I get along without you I want to lay my head on your breast as I did before oh so much this morning. I went to bed last night the first time since I left you. (This suggests that he traveled across the country by train, sitting up all the way, which was common in that decade.) I was very tired and went to sleep quick but this morning where is Addie.

"I will say a good morning to you and Cora. RDW"

Later that day he wrote again,

"I hope I shall make enough this summer to be able to go to my dear good kind patient loveing wife Addie. I love you as a drowning man loves the sight of a sail. I feel today just as if you were the only being on earth that cared for me and tis no doubt true.

"I have been to have some pictures taken to please you. I did not want to and besides they cost $4.50 but if they do please you it is cheap enough. I shall send them to you when they are done. I shall not see them but I have got the best I could. now about your Home. . . ."

The details of what was apparently a complicated family quarrel involving the deed to their New Bedford home follow, at the end of which he advises,

"if you must, take your things and Cora and Hire a House in New Bedford or wherever you want to, you have enough to keep you in either case until I come to you again.

"I should feel very badly about this Addie if I did not know you so well. I know it may be very unpleasant for you but I know nobody will get to windward of you anyway.

"I have been down to the Office and Shipped and we sail tomorrow morning so I doubt if I write again. . . I am afraid I shall not get Cora's picture and yours, how much I wanted yours Addie. I have got one but I am afraid to look at it yet, I am very foolish that way."

Not only did Cora's picture fail to reach him before he left; he was not to see his first daughter, whom Addie had been taking care of, again. The young girl died of consumption in August, 1885, at the age of thirteen, and was buried beside her mother in Falmouth, Massachusetts.

Helen's birth about a year later was thus doubly treasured by Captain and Mrs. Wicks, and it is perhaps because of all the sadness which had preceded it, added to the fact that she was a woman of spirit who loved her husband, that Adelaide agreed, when the child was two, to make the long rail journey across the United States and embark on her first whaling voyage.

After the accident near Rock Springs, Wyoming, she tells of stopping at Ogden, Utah, and at Carlin, Humboldt and Wadsworth, Nevada--"very high stone cliffs, barren soil, and many Indians, men and some squaws" --before they reached California on November 18.

"Awoke and rose at 4-10. Sacramento, some passengers left the train, rainy weather with occasional glimpses of the moon. Dolph met me at 16th Street, Oakland."

For the next few days the captain and his wife "bought table linen and a looking glass and books" for the ship, wrote innumerable letters back to New Bedford, consumed gallons of ice cream – a luxury which a whaling bark would not provide --and finally, on December 4,

"went to the ship on the tug Alert, went on board ship at 10-30, fair wind, was seasick at 3."

By December 6 she was well enough to eat "steak and boiled potatoes and toast for breakfast" in her cabin, watch her husband Dolph build a small berth for Helen, and to begin the routine which was about all life aboard a whaler offered a 19th Century woman: reading, sewing, crocheting, amusing her child, napping, walking or sitting on deck, laundry, eating, and recovering from a hazard common among families who traveled on such drafty voyages: bouts with the common cold.

On January 4, about a month out of San Francisco, the Coral "caught 4 sperm whales which were cut in the next day it was very rugged stored down 62 bbls Oil. A sail has been seen twice in the distance, it has been very hot and showery for days. Dolph and Helen are sitting here sewing, there comes a call from the mast head 'Blow, blow, blow,' Dolph runs and Helen and I go below."

Sunday, February 10 found the Coral in the Marquesas Islands "off Magdalena waiting to anchor, it is very hot here indeed. the inhabitants are natives under French government, the land is very high, 3,680 feet. cotton is raised here, Bananas, Oranges, Pineapples, Mummy apples, cocnuts, Breadfruits.

"One of the men sent the Capt's wife a live chicken and some eggs because she had been sick. the chick is walking around the deck eating coconut.

"two weeks ago tomorrow I had a severe attack of pain in my left side it confined me to bed several days after that I began to improve."

Her "first storm at sea" came a few days later, when "we experienced a sudden and severe Squall of wind and rain. Saw the ocean in all its wild Grandeur then laid on the sofa and thought of Ma and Home."

The tempest passed, they put in at another torrid island for "13 casks of water," and Mrs. Wicks adds, "it is a very wild country, the natives come on board the ship very often, they also cut wood for us then swim with it to the boats and our men take it in, we can see them from the ship with a glass, the most of it is Rosewood.

"the natives have a great deal of tatooing on them some of it is very finely done, they wear scarcely any clothing. we went ashore for a few hours, visited the cemetery, went to the Governors, drank coconut, saw the native women and children. Helen was quite a sight to them, she said good morning and it seemed to please them very much. . . .

"the Governor and his wife took dinner with us on Thursday on board the ship she was dressed in white with a train a white flower over her ear and barefoot."

The next days, at sea once more, were "rainy and hot, very light or no wind, Helen trotting around with a calico cap on her head, several sharks have been killed also one Porpoise and two skip jacks the latter taste something like our Swordfish."

Rain fell "every day for nearly two weeks" as they headed back toward Honolulu, where Adelaide and Helen would disembark. "very rough weather, rolling and tumbling about," she notes until on March 17,

"Off the island of Hawaii in a dead calm, we sighted this island yesterday it was very rugged last night we were becalmed. . . can see the smoke from the Volcano resting in the valley also a church and other buildings. this morning heard one of the men singing 'Lily of the Valley' it sounded natural and good.

"a week ago one of the Boatsteerers met with an accident a hatch fell on his foot causing a severe sprain he is able to be out today, another man has been sick with severe attacks of fits so that he has had to be chained."

Three days later the Coral was "still drifting, I have been making some paper flowers." Four more days went by before "a breeze has sprung up and we are going along quite glibly, three other islands of the group are in sight it is very pleasant and cool." In other words, Hawaii had been visible for some eight days before they could land — a possible thought-provoker for impatient 20th Century air travelers.

Ashore in Honolulu the Wicks family "engaged rooms and board at the Eagle House, Nuuana Avenue very pleasant place flowers birds and pets in abundance," drove in a hired carriage "to the Park, did some shopping bought Helen hat, hose, shoes, some bracelets, called on Miss Hattie Pierce, Mrs. White called, left her card, got some ice cream in the evening." During the next days "we went to see the tomb where the Kings are buried, saw rice growing, saw and tasted poa," paid and received a half dozen more calls, "bought H. 2 dolls, Pongee dress, card case, Crepe and silk hankerchief."

On the day of the last purchases, March 25, Adelaide and Helen's luggage was stowed aboard the U.S. Bowne, a cargo vessel "loaded with raw sugar, the odor made us quamish," for San Francisco.

Captain Wicks, bound now for a longer voyage to the Arctic in the Coral, escorted his wife and daughter to the Bowne, then "went back with the pilot and we had a grand salute from the Coral as we passed her."

The trip from Honolulu to San Francisco took eighteen rough days, during which the vessel developed a leak and Adelaide comments, "the captain is worried about the sugar getting wet, nothing more, it is worth $60,000."

She and Helen returned to New Bedford sometime in the spring of 1889 although the journal, which ends on the West Coast in April, does not say when. Captain Wicks must have re-joined them by late fall, for a second daughter, Lillian Frances, was born in August of 1890.

The captain himself was born in West Falmouth in 1846, signed his first papers at seventeen to go to sea on the Bark Sarah out of Mattapoisett, and received his license as master seven years later — one of the few to acquire this rank by the age of twenty-four. He was captain in subsequent years not only of the Sea Breeze and Coral, but of the Mermaid, and the steam whaler Beluga.

Because one of his many other voyages was made aboard the Lagoda, whose half-scale model, the largest replica of a whaler in the world, is a featured exhibit in the Whaling Museum of the Old Dartmouth Historical Society, he was consulted about many of the details of the model's rigging when it was built in 1915 and 1916.

Captain Wicks was seventy-nine when he died in New Bedford in March, 1925.

The bark Coral, photographed by Albert Cook Church during rebuilding, took Captain and Mrs. Rodolphus Delano Wicks and their two-year-old daughter Helen on a whaling cruise to the South Pacific in 1888 and 1889.

About the Descendants of Whaling Masters

The Descendants of Whaling Masters, Inc., formed in 1974 in New Bedford, Massachusetts, has at this writing two hundred and twenty members from fourteen of the United States.

Its goals are to perpetuate the history and traditions of the whaling era; to encourage members to preserve their whaling artifacts, records, letters, and journals, and to provide for their safe keeping; to inform young people of the far-reaching significance of the whaling period, including its impact on the economy; and to provide opportunity to members and friends for the sharing of common interests.

This volume has been edited and compiled by Genevieve M. Darden, a member of DWM and an associate professor in the Department of English, Southeastern Massachusetts University.

March, 1980